# NUCLEAR
# ETHICS

# NUCLEAR ETHICS

Joseph S. Nye, Jr.

**THE FREE PRESS**
*A Division of Macmillan, Inc.*
NEW YORK

Collier Macmillan Publishers
LONDON

The Free Press
A Division of Macmillan, Inc.
866 Third Avenue, New York, N.Y. 10022

Collier Macmillan Canada, Inc.

First Free Press Paperback Edition 1988

Printed in the United States of America

printing number

1   2   3   4   5   6   7   8   9   10

**Library of Congress Cataloging-in-Publication Data**

Nye, Joseph S.
   Nuclear ethics.

   Bibliography: p.
   Includes index.
   1. Nuclear warfare—Moral and ethical aspects.
2. Deterrence (Strategy)—Moral and ethical aspects.
I. Title.
U263.N753   1986        172′.42        85-29373
ISBN 0-02-923091-8

*For my parents who taught me values*
*from past generations*
*and*
*For John, Ben, Dan*
*and their future generations.*

# Contents

# Preface and Acknowledgments

THE PROSPECT of a nuclear war is horrifying. It brings us face to face not only with death, but with destruction of the civilization that makes our life meaningful. It might even destroy our species. There is no precedent for the challenge that nuclear weapons present to our physical and moral lives.

Each of us is inescapably involved in the nuclear dilemma. At the very least, we are targets and victims. As voters and taxpayers in a democracy, we are also participants in the system of defense by nuclear deterrence. Many people, including some nuclear strategists, prefer to avoid or conceal the questions of conscience that arise from this situation. But many others realize that refusal to think about choices among values in such situations is itself a form of choice.

In 1976 I visited Hiroshima. Anyone who thinks of nuclear weapons as abstractions should visit the Hiroshima Museum. No one can help but be moved by the real human horror such weapons can produce. My Japanese guide was a survivor. I asked him how he felt about the American action three decades later. He confessed that Japan would have done the same thing if it had had the bomb first. What he could not understand was why so many innocent people had to die. His question has haunted me. Subsequently when I worked in the State Department on efforts to stem the spread of nuclear weapons, I traveled to countries like India, Pakistan, Argentina, and Brazil, where I was asked why it was right for Americans to have nuclear weapons and other countries should not. That question also troubled me. Later, when I returned to my teaching at Harvard, a number of my students would ask me why we needed nuclear weapons and whether we were doomed to perish in a nuclear war.

I have written this little book in part to answer those questions for myself—to help me think through the moral challenge of a subject with which I have been deeply involved for the past decade. I have also written it for others who are concerned with the same questions. I have tried to keep the style simple and the argument clear. The book is addressed to citizens rather than specialists in strategy or philosophy, though the arguments must pass scrutiny by such specialists.

Two problems plague efforts to think clearly about nuclear ethics. What is morality? Does it apply to relations between nations? I regard the answers I have given to both questions in the first half of the book as equally important as the specific recommendations about nuclear weapons in the latter part of the book. This is not really a book about nuclear weapons; it is a book about how citizens in a democracy can carry on moral discourse on a subject that sometimes seems to pose an intractable challenge to our most basic values.

Sometimes people regard morality as an expression of unexamined and unchallengeable first premises or basic intuitions. A moral position, in that view, is an emotional expression that tries to move other people's intuitions. The correct moral response to nuclear weapons is thus a cry of moral outrage. But whatever the role of moral outrage, it is not the same thing as moral reasoning, and hardly a complete description of our ethical traditions.

Others regard morality not as outrage but as rigid application of a set of rules. Any effort to calculate consequences or give reasons for waiving rules in certain circumstances is regarded as mere pragmatic reasoning, not moral reasoning. But as I shall explain, that is a stunted view of moral reasoning which ignores a large part of the great Western traditions in moral philosophy.

I shall argue that the morality of nuclear deterrence is conditional. Deterrence as a national strategy cannot be absolutely justified or condemned without further evidence about motivations and likely results. Establishing such conditions is critical for moral reasoning about nuclear weapons. I have searched for such conditions by examining the nature of both nuclear weapons and world politics, and by appealing to principles and rules that have stood the test of history. I have drawn from the just war doctrine of the past to suggest a "just defense doctrine" for the nuclear age. I believe that such a doctrine can provide a moral compass for choices as well as a sense of hope for the future. Without such hope, we may fall prey to a cynicism and nihilism that are deeply corrosive to our lives as moral beings.

The method of moral reasoning about nuclear weapons applied here does not invoke outrage, but it does encourage a strong moral commitment to avoiding nuclear war. Commitment is not the same as hysteria. And crusades based on moral outrage can lead to horrendously immoral consequences. Morality is not the opposite of reasoning. Even though human reason has its limits, we are not excused from using that frail but crucial instrument in our application of deeply held values. Otherwise a complex world would constantly ambush our best intentions.

This book is not the end of wisdom or argument about nuclear ethics. It is merely an effort to clarify and promote a dialogue that has recently been bedevilled by shrill invective and by destructive pessimism about the future. Open dialogue is the path to progress in moral reasoning in a democracy. And serious dialogue is the path to the refinement of ideas.

In that respect, I am grateful to a large number of people and institutions for their help. This book is a joint publication of the Center for International Affairs and the Avoiding Nuclear War Project of the Center for Science and International Affairs at Harvard University. The latter project is supported by the Carnegie Corporation of New York. A small grant from the Ford Foundation to the Aspen Institute made possible two small discussion meetings in the early stages of the project. And a month's sojourn as a scholar in residence at the Rockefeller Foundation's Villa Serbelloni provided a fruitful setting for writing near the end. I am grateful for the support of those institutions and their staffs.

For helpful comments on earlier formulations of my argument, I am indebted to: Graham Allison, James Blight, Sissela Bok, Harvey Brooks, McGeorge Bundy, George Bunn, Albert Carnesale, Ashton Carter, Richard Cooper, John Deutch, Tom Donaldson, Michael Doyle, Charles Glaser, Thomas Graham, Russell Hardin, Brian Hehir, Stanley Hoffmann, Stephen Holmes, Carey Joynt,

Steven Kelman, William Kristol, John Langan, Sean Lynn-Jones, Theodore Marmor, Ernest May, William Maynes, Walter McDougal, Jim Miller, Terry Nardin, Uwe Nerlich, Michael Novak, Susan Okin, M. J. Peterson, Ralph Potter, James Rhodes, Scott Sagan, Michael Sandel, Thomas Schelling, Enid Schoettle, Sanford Segal, Judith Shklar, Henry Shue, Harrison Wagner, Michael Walzer, and Richard Zeckhauser. Robert Keohane deserves double thanks, since he suffered through the whole manuscript twice. John Wertheimer deserves thanks for help with footnotes and index.

Molly Harding Nye not only read the manuscript, she lived it with me. As with so much, the dedication to John, Ben, and Dan and their future generations is from both of us.

# CHAPTER ONE

---

# Thinking About the First Generation

A SINGLE BOMB destroyed Hiroshima. By current standards it was a small nuclear weapon. A single three-megaton hydrogen bomb equals all the conventional bombs dropped in World War II. Today there are some 50,000 nuclear weapons in the world. Their full use in war would destroy our civilization. A fraction used against cities could possibly contaminate such a large portion of the earth's atmosphere that all human life would cease, at least in the Northern Hemisphere. We may be, in the words of the American Catholic Bishops, "the first generation since Genesis with the capacity to destroy God's Creation."[1]

What does this unprecedented situation mean to our moral lives as human beings? Is it true that "as long as

1

there are nuclear weapons in the world we are compelled to choose between a position that is politically sound but immoral and one that is morally sound but politically irrelevant?" Do "we incur the full burden of guilt for extinguishing our species merely by preparing to do the deed, even without actually pushing the button?"[2] Have nuclear weapons "exploded" the social conventions, such as the just war doctrine, that we have inherited from the past?[3] Can we morally justify the possession of such weapons?

## Public Views

If moral argument were merely a game that theologians and philosophers played in their cloisters, practical politicians might feel tempted to ignore such questions. But those issues have increasingly concerned the public in democratic societies. There has been a spate of books, articles, films, and television programs on nuclear issues in both Europe and the United States in recent years. Prominent churchmen have issued statements on the moral issues of nuclear deterrence. Government leaders as well as their political opponents have used the language of morality. Our nuclear arsenal is being "shaken by a war of ideas about its purpose."[4]

The current period is not the first time that public opinion in the democracies has been aroused over nuclear issues. A similar concern marked the period from 1958 to 1962. It ebbed with the ensuing improvement of U.S.–Soviet relations and the beginning of arms control agreements. Concern returned with the deterioration of relations in the late 1970s. Both periods were marked by a heightened sense of the prospect of nuclear war. Like the present, the earlier period also was marked by public debate over the morality of nuclear weapons and nuclear deterrence. Most of the basic moral questions remain unchanged.

Some dimensions of the current situation are new. The number of strategic nuclear weapons has increased, though the total destructive power—"equivalent megatonnage"—of the American arsenal is less than in 1962. Technology has made both nuclear and conventional weapons more accurate and has reduced the danger of some kinds of accidents. Nuclear weapons have spread to several more countries. American and Soviet leaders have more than two decades' additional experience of deterrence without war, including the surmounting of nuclear crises in Cuba and the Middle East. But the moral dilemmas persist.

The American public has reacted to this situation in ambivalent ways. Half think more about nuclear war than they did five years ago; half say the whole idea is so horrible that they try not to think about it. Most (89 percent) believe there can be no winner in an all-out nuclear war and that nothing on earth could ever justify the all-out use of nuclear weapons (79 percent). But only a third say they would never use nuclear weapons under any conditions; slightly more (41 percent) say they would rather die in a nuclear war than live under communism. A quarter of the public believes that in a nuclear war with the communists, our faith in God would ensure our survival, but only a fifth (21 percent) of the public expect personally to survive a nuclear war. A large majority (79 percent) rejects the idea that we should have used nuclear weapons against the Soviets before they developed nuclear weapons of their own. In general, there is less optimism about nuclear weapons today than existed three decades ago. In 1949 a considerable majority (59 percent) of the public thought it was a good thing that the atomic bomb was developed; by 1982 nearly two-thirds (65 percent) thought it a bad thing.[5]

Individual responses vary from apocalyptic vision to complete cynicism. Some religious fundamentalists see nuclear war fatalistically as the inevitable Armageddon predicted in the Bible; our choices are largely irrelevant to what is foreordained.[6] At the other extreme, cynics also regard moral argument as largely irrelevant. As the cynical senator replied to the Sunday school teacher's question about whether morality played any role in politics, "Of course it does, you use any weapon you can in a political fight." According to the cynics, first one picks a nuclear policy, and then one invents some moral garments to cover it.

There is plentiful evidence of such instrumental use of moral arguments in the nuclear debate over the years. For example, faced with budget cuts in 1949, "the Navy attacked the A-bomb, then sole property of the U.S. Air Force" as immoral. "Then in 1951, the Navy started to assemble its own atomic arsenal. Suddenly, only a few years after, . . . the atomic bomb was no longer so 'barbaric.' "[7] During the 1970s some critics of the strategic doctrine of mutual assured destruction (MAD) attacked it as immoral because it implied the destruction of innocent civilians in cities, but some of the same critics favored nuclear attacks on leadership, industry, and communications centers, which would be equally disastrous for civilians in cities. Similarly in the current debate over building a strategic defense against ballistic missiles, grandiose moral arguments are loosely used. Some proponents call it a "moral obligation." They attack the immorality of the strategic doctrine of assured destruction long after MAD ceased to be official doctrine rather than address the harder question of whether one can escape the *condition* of MAD without raising the likelihood of nuclear war. And some opponents assert categorically that the effort to develop strategic defense is immoral even though its dimensions are still largely unknown.

But cynics jump too quickly to the conclusion that moral reasoning is merely an irrelevant fig leaf used to cover up reality. Even in the instrumental sense, poor moral reasoning fails to move the minds and consciences of fellow citizens. The fig leaf is too small or the garment too tattered. The inconsistency of the Navy's argument made it too transparent to be compelling. In short, the quality of moral argument makes a difference, and not merely among "doves." Hawkish strategists recognize that "the ability of a democracy to sustain an adequate military posture, year after year (for decades and even longer), is not unrelated to the popularly perceived compatibility of moral values with defense policy."[8] That consideration is doubly important for a country like the United States, which has defined its security in terms of an alliance of democracies.

The case for careful attention to moral reasoning, however, is not merely instrumental. Ethical considerations often move people, at least in part. Most leaders do not live wholly by the word, but neither do they live solely by the sword. Mixed motives are a fact of human life. It is as mistaken to dismiss the ethical dimension as it would be to ignore the pragmatic aspect in the motives of leaders. For example, ethical considerations played a role in Henry Stimson's removal of the Japanese cultural center, Kyoto, from the target list for the first atomic bomb; in the American rejection of a preventive nuclear war against the Soviet Union in the early 1950s; and in John Kennedy's choices during the Cuban missile crisis of 1962. Even President Eisenhower, who sometimes talked about nuclear weapons as normal military weapons, told his advisers who suggested the limited use of nuclear weapons to prevent the fall of Dien Bien Phu in 1954, "You boys must be crazy. We can't use those awful things against Asians for the second time in less than ten years. My God!"[9]

Moral considerations will probably affect any statesman or soldier who ever faces the prospect of using nuclear weapons. For example, Admiral James Watkins, Chief of Naval Operations and a practicing Catholic, has said he regards the first use of nuclear weapons as "a very significant problem for me from a moral standpoint." Asked what his moral criteria for employing nuclear weapons would be, Admiral Watkins said: "I would have to know the entire scenario up to this point. How did we get into this situation? What alternatives do we have? Have I used up every single alternative at my fingertips? Are we about to see the demise of everything we cherish? Are we about to lose the Western world and democracy? Is it very clear that it is now a question of subservience for an undefined period of years? Have I attempted to negotiate with the Soviet Union with the most powerful tools that I have left? Have we reached a stage of hopelessness? Those are the kinds of things that would go through my mind. It would have to be that hard."[10]

## The Skeptic's Escape

Some thinkers have tried to avoid the problem of thinking about the ethical aspects of nuclear strategy by exempting the whole domain of international politics from moral considerations. Unlike the cynics, skeptics about international politics accept the significance of ethics in domestic politics but argue that moral categories and judgments have no meaning in relations among nations.[11] In the absence of strong global institutions and a sense of international community, politics among nations remains a jungle. As Thucydides put it two and a half millenia ago, "the strong do as they will and the weak suffer what they must." Self-interest is more important than justice in such a setting. *Raison d'état,* or the self-interest of the state, is the sole criterion for policy choices. Prudence in the pursuit of self-interest is wise, since obviously nuclear self-destruction would be bad. But costs imposed on noncombatants or on future generations or on nonnuclear nations are not a moral concern. They are simply the unfortunate products of necessity in a totally amoral domain. To be prudent in the pursuit of your interests is sufficient. It is not necessary to consider the interests of other nations. As a French diplomat once told me, "Since there can be no certainty in international morality, the only sound position is the interest of France." There are no ethical issues to consider in nuclear strategy; there is only enlightened self-interest.

Considering the nature of the international milieu, it is not surprising many diplomats and serious students of international politics have tended to be cautious about the role of ethics in foreign policy and have warned about the possible disastrous consequences of well-intentioned moral crusades in such a difficult domain. But there is a difference between healthy realism and total skepticism. It does not follow from the difficulty of applying ethical considerations that they have no role at all. The total skeptic who argues that there is no role for ethics in international politics tends to smuggle his preferred values into foreign policy, often in the form of narrow nationalism. When faced with moral choices, to pretend not to choose is merely a disguised form of choice.

Philosophers often say that "ought" implies "can." When something is impossible, we have no moral obligation to do it. If international politics (including the nuclear balance) were the domain of absolute necessity, then there would be no room for morality. But absolute necessity is extremely rare. Even in situations of acute insecurity there are often choices with profound moral implications. A decision to use nuclear weapons would be such a choice. Moreover, many choices arise before

7

such moments of ultimate peril. One cannot legitimately banish ethics simply by asserting that international politics is a "state of war" or that we are engaged in a "cold war" with an amoral adversary. We do have choices about nuclear weapons, about our foreign policy interests, and about how we pursue them. As Arnold Wolfers pointed out, much of international politics allows choices about the definition of indefinite concepts like "national interest," "survival," and "prudence."[12] The statesman who says, "I had no choice," usually did have, albeit unpleasant ones!

In their personal lives, most people feel strongly bound by the sixth commandment: "Thou shalt not kill." But presidents may have to take decisions that violate that rule if they are to protect their own people in wartime. The fact that international politics is a difficult domain for ethics means that one must be cautious about too simple a transposition of moral maxims from relations among individuals to the domain of states. But being president does not release the statesman from the duty of moral reasoning; it merely complicates his or her task.[13] One must examine the arguments leaders give for claiming there is no choice or for why they think normal moral rules that we use in daily life should not be applied in particular cases. The burden of proof rests on those who wish to depart from normal morality. While that burden may often be met, the quality of their argument and conclusions deserves close examination. Some arguments for disregarding normal moral rules are fallacious.

For example, the fact that nations must act to defend their interests if they wish to survive in international politics does not mean that only selfish acts are possible. It does mean that most international acts will involve mixed motives. But some degree of altruism or consideration of others can often be included in the motives for foreign policy. Nor does the existence of different national moral standards make our own moral choices impossible. The sociological fact that different cultures have different values should makes us cautious about the potentially immoral consequences of blindly trying to impose our moral standards on other cultures. But it does not excuse us from making moral choices about our own actions. If Iraq uses poison gas against Iran, or if the Soviets execute prisoners in Afghanistan, it does not make it right for us to do so. Two wrongs don't make *us* right!

Some people question whether the United States can afford to act ethically when it is locked in a bipolar rivalry with a Soviet adversary whose doctrine rejects "bourgeois morality" and sees the goal of proletarian victory as justifying the means they use. The difference in moral views should alert us not to expect the Soviet Union necessarily to behave as we do, but it does not justify our behaving as they do. We often choose to act morally because of our desire to preserve our integrity as a people. There are certain things we just don't do because of the shame they would make us feel about ourselves. As James Schlesinger has put it, "democracies forgo certain options by the nature of their societies and the whole set of ideals they represent."[14] To ignore Soviet behavior in our strategic interaction would be foolish, but to imitate it slavishly would be a particularly insidious way of losing our sense of moral integrity as a people and a society. If we possess nuclear weapons to help us to defend our values, it is important to remember that there is more than one way to lose our values.

Some skeptics question the application of moral concepts to individuals who live beyond our borders such as civilians in the Soviet Union or citizens of third countries. Where political processes and communities are separate, why should we be concerned about just treatment of people who live beyond our borders? I shall examine that question more carefully in Chapter 3, but to the

9

extent that we accept responsibility for the effects of our actions on others, we have to realize that the effects of nuclear weapons are almost inevitably transnational in their scope. Their very nature calls into question the limited perspective of the total skeptic. The skeptics are correct that nations must defend themselves in a world that acknowledges no higher government. But a right of self-defense does not entail an unlimited right with no concern about costs imposed on others. The skeptic about international relations does not provide a compelling excuse for avoiding moral reasoning about nuclear weapons. He merely ducks hard questions about why he should treat his nation as his only international value.

## The Tone of Moral Debate

Moral reasoning about nuclear weapons is inescapable in democracies. It is an unavoidable part of the human condition of the "first generation since Genesis." Though the subject is bound to be uncomfortable, and the arguments sharp, the debate can be improved by the practice of the ordinary virtues of humility and charity. Humility is needed because it is unlikely that any argument will satisfy fully all the deeply held values at stake. And self-righteousness blinds people to what is valid in the perspectives of others. There is an important difference between self-righteous moralizing and careful moral reasoning.

Humility has been notably absent in the nuclear debate. Many strategists ignore nuclear ethics, and many moral absolutists refuse not only to tolerate nuclear weapons, but even to tolerate those who do, preferring to categorize strategists and their arguments as corrupted. All too often moralists and strategists tend to talk past each other as though they lived in separate cultures of warriors and victims rather than fellow citizens of a democracy.[15] The moralists formulate fine principles that seem to the strategists about as relevant to a foreign policy as a belief in the tooth fairy is to the practice of dentistry. The strategists, on the other hand, tend to live in an esoteric world of abstract calculations and a belief in a mystical religion called deterrence, which is invoked to justify whatever is convenient. Strategists would do well to realize that there are no experts, only specialists, on the subject of nuclear war, and to listen more carefully to the moralists' criticisms. At the same time philosophers and moralists would do well to pay more heed to the strategists' arguments and to realize that they will need to work with more realistic assumptions if they wish to be effective in a dialogue between ethics and strategy.

Even when strategists do consider moral principles and moralists do pay attention to consequences, charity is needed if moral arguments are to be heard. Shrill charges of immorality ("no one committed to defence by nuclear weapons can have any principled objection to murder")[16] and unrealism rarely advance moral reasoning. Snide attributions of bad motives quickly stifle dialogue. It is hard to find any leaders who want a war that will end the species, though some are treated as though they did.[17] Sometimes there are differences over values, but often there are quite different causal assumptions and judgments about acceptable risk that are at issue. Not all can be resolved. Some involve deep differences of values, but some involve faulty moral reasoning about such differences.

Some who oppose nuclear deterrence discount the views of those who defend it as corrupted by "the disease of nuclearism."[18] Instead of meeting their opponents' arguments, they make up a theory about their opponents' motives. They try to shrink their opponents' stature rather than refute their arguments. When they do that, they are involved in political caricature, not in moral reasoning. Without a degree of humility and charity, we are condemned to shouting such caricatures at each other, and the illumination of moral reasoning is snuffed out. Sometimes such exaggeration is defended because it is expected to have positive political effects. Shouting may help to attract political attention; it rarely serves to keep it for prolonged periods. Other people believe that exaggeration is justifiable moral outrage in the face of an extreme moral dilemma. They argue that outrage and exaggeration are needed to change our way of thinking.[19] The hortatory or emotional style of the cru-sader or prophet is more appropriate than the rational style of the philosopher if the task is to awaken moral revulsion and change moral perceptions. As an activist friend once told me, he sees his task as "smearing" nuclear weapons so that the public loathes them as much as it does chemical weapons.

Sometimes that emotivist approach to nuclear ethics is bolstered by quoting Einstein: "the splitting of the atom has changed everything except our way of thinking, and thus we drift towards unparalleled catastrophe." Reasonable responses seem inadequate to the challenge. "Reinventing ourselves means changing our minds, as Einstein suggested we must."[20] But who are "we"? If humankind is currently divided in a manner that permits nuclear weapons to be smeared or delegitimized only in democracies but not in other parts of the world, the emotivist approach may make the use of nuclear weapons as well as the loss of our values more rather than less likely. Nuclear education is necessary, but it must make the world better, not worse, if it is to have moral effects; and that requires careful attention to facts and reasoning about the way the world is as well as how it ought to be. There is a crucial difference between moral outrage and moral reasoning. Outrage generally prevents reasoning and sometimes leads to disastrous consequences. It is hardly the new form of thinking that is called for by our nuclear peril.

Let us hope that outrage and the abandonment of moral reasoning are not the human condition of the "first generation since Genesis"! Our condition is too serious to leave to amoral strategists, moral absolutists, or apocalyptic activists. What an ironic comment on the human condition if our only form of moral discourse about our nuclear predicament would be to shout insults at each other.

I discuss the methods of moral reasoning in the next chapter and ask how we are to judge good from flawed moral arguments. Chapter 3 turns to the scope of moral reasoning and the question of what obligations we owe to foreigners. Those impatient with moral philosophy may wish to skip those chapters, but I hope not. All too often, moral arguments about nuclear weapons (as well as other issues in foreign policy) show too little care about their undergirding philosophical foundations. Chapter 4 discusses the difficult problems of ends and means that nuclear weapons raise for just war theory. Chapter 5 looks at the consequences of nuclear deterrence; the question of whether a disastrous failure is inevitable; and what obligations we owe to future generations. Chapter 6 is concerned with the effects of nuclear weapons on third countries that do not have such weapons (but in some cases are trying to get them.) Finally, the concluding chapter suggests five maxims for applying moral reasoning to nuclear weapons in a way that meets our obligations to the present and future generations.

# CHAPTER TWO

---

# How to Judge Moral Reasoning

$W$E WISH to encourage good moral reasoning about our nuclear predicament, but how do we judge moral reasoning? Some people think there is little to judge, since ultimate ends are not susceptible to a common proof in modern Western culture. It is true that our culture seems ethically fragmented compared to the world of the ancient Greeks, the Medieval Church, or a remote tribe.[21] But that simply means our moral reasoning will not be tidy.[22] It does not prevent us from engaging in moral reasoning or from reaching common judgments about many moral arguments. In fact, like Moliere's character who discovered he had been speaking prose all his life, so have we been judging moral reasoning every day despite our alleged inability to agree upon ultimate ends.

Much of moral argument is not about ultimate ends but concerns means, and expected consequences, and the relationship between them. And even deeply held views of ultimate ends can be refined, if discussion goes beyond primitive assertion. Suppose one person asserts that nuclear weapons are good and another says they are evil. If the only reasons they give are their intuition or the revealed truth of their religion, there may be little more the two can say to each other. If all they can say is "I just feel it," they are making a very stunted moral argument. But such primitive assertion is relatively rare. More important, it is rarely compelling to others who do not already share that particular intuition or that same source of revealed truth. If we wish our moral arguments to be compelling to our fellow citizens, we need to go beyond primitive assertion.

In fact, we constantly judge moral arguments in terms of their clarity, logic, consistency, and unnoticed negative consequences.[23] For example, if someone argued that nuclear weapons are evil because they demonstrate "scientific hubris," one could ask whether that fuzzy term does not also condemn kidney transplants and much else in modern medicine. Or when Phyllis Schlafly argues that nuclear weapons are good because God gave them to us,[24] one can ask why He allowed the atheistic Soviet Union also to get them. Or if it is argued that nuclear weapons are good because "the bomb is modifying people's thinking. . . . the bomb is pointing to the problems of the world globally,"[25] one can ask if there are not less dangerous ways to pursue such understanding. Or take Jonathan Schell's argument that "although, scientifically speaking, there is all the difference in the world between the mere possibility that a holocaust will bring about extinction and the certainty of it, morally they are the same." Others reply that to blur deliberately the distinction between a scientific certainty and an unknown prob-

15

ability "is evidence of intellectual hysteria."[26] We are constantly examining moral arguments for their clarity, logic, consistency, and unnoticed consequences. Those which fail such tests tend to be less compelling.

## Two Ethical Traditions

An ethical judgment ("murder is evil") is distinguished from a mere expression of taste ("pizza is good") by combining prescription (it tells you what to do); overridingness (it tells you what is important); and impartiality (which is illustrated by the golden rule: Do unto others as you would have them do unto you).[27] Some degree of impartiality is generally regarded as a touchstone of ethical thinking. "What distinguishes ethical principles is the disinterested perspective they embody. Prudence asks whether an action or policy serves the interest of some particular individual or group or nation. Ethics asks whether an action or policy could be accepted by anyone who did not know his or her particular circumstances (such as social class, race, or nationality)."[28] But what degree of impartiality and which impartial standards to use are contentious subjects.

Within the Western ethical tradition, significant differences exist, particularly over the role of personal virtue versus the effects of actions. The two approaches might be called the ethics of virtue versus the ethics of consequences.[29] The first approach focuses on the quality of the person doing the act, and the second focuses on the consequences of the act.[30] The consequentialist tradition places its emphasis on the outcomes of actions. Utilitarian philosophers like Jeremy Bentham or John Stuart Mill, who try to calculate which acts do the greatest good for the greatest number, are in the consequentialist tradition. The person-centered approaches stress whether a person is following rules and has the right motives as the basis for judging the morality of actions. The famous German philosopher Immanuel Kant has come to symbolize the latter approach. The difference between the two moral traditions could be described as the difference between an emphasis on my integrity judged in terms of whether my actions conform to certain rules and an emphasis on the consequences of what I have done regardless of my motives. One focuses more on the actor's process of deciding and the other on the effect of the decision.[31] Both traditions express important truths.

Philosophers who stress personal integrity argue that we often refuse to do things that could have good consequences because such actions would violate a universal moral rule or another person's absolute right and would thus damage our own moral integrity. For example, if we believed that a robbery suspect was innocent but that his imprisonment would have a beneficial effect of deterring others, the beneficial deterrent would not justify our sending him to jail. It might do "good," but it would not be "right." In fact, our whole judicial system goes to great lengths to protect the rights of those accused of crimes in order that we not unjustly punish an innocent person. The tradition stresses that it is wrong to do evil as a means to do good. The moral person would rather suffer an injustice than commit one.[32]

Such philosophers apply those principles to nuclear deterrence. It is wrong to threaten innocent life even if it may help deter war. How can a country live with its conscience and know that it is prepared to kill twenty million children in another country if worse should come to worst? In a haunting analogy, the Protestant theologian Paul Ramsey likens the targeting of cities to placing babies on the bumpers of automobiles to make holiday drivers more prudent. It might save lives, but we would be wrong to do it.[33]

The significance and the limits of the two broad traditions can be captured by contemplating a hypothetical case.[34] Imagine that you are visiting a Central American country and you happen upon a village square where an army captain is about to order his men to shoot two peasants lined up against a wall. When you ask the reason, you are told someone in this village shot at the captain's men last night. When you object to the killing of possibly innocent people, you are told that civil wars do not permit moral niceties. Just to prove the point that we all have dirty hands in such situations, the captain hands you a rifle and tells you that if you will shoot one peasant, he will free the other. Otherwise both die. He warns you not to try any tricks because his men have their guns trained on you. Will you shoot one person with the consequences of saving one, or will you allow both to die but preserve your moral integrity by refusing to play his dirty game?

The point of the story is to show the value and limits of both traditions. Integrity is clearly an important value, and many of us would refuse to shoot. But at what point does the principle of not taking an innocent life collapse before the consequentialist burden? Would it matter if there were twenty or 1,000 peasants to be saved? What if killing or torturing one innocent person could save a city of 10 million persons from a terrorists' nuclear device? At some point does not integrity become the ultimate egoism of fastidious self-righteousness in which the purity of the self is more important than the lives of countless others? Is it not better to follow a consequentialist approach, admit remorse or regret over the immoral means, but justify the action by the consequences? Do absolutist approaches to integrity become self-contradictory in a world of nuclear weapons? "Do what is right though the world should perish" was a difficult principle even when Kant expounded it in the eighteenth century, and there is some evidence that he did not mean it to be taken literally even then. Now that it may be literally possible in the nuclear age, it seems more than ever to be self-contradictory.[35] Absolutist ethics bear a heavier burden of proof in the nuclear age than ever before.

On the other hand, the dangers of too simple an application of consequentialism are well known. Once the ends justify the means, the dangers of slipping into a morality of convenience greatly increase. To calculate all the consequences of one's actions is impossible, and when the calculations are fuzzy, abuse is likely. The utilitarian who tries to judge each act without the benefit of rules may find the task impossible to accomplish except with a shallowness that makes a travesty of moral judgment. And given human proclivities to weight choices in our own favor and the difficulties of being sure of consequences of complex activities, impartiality may be easily lost in the absence of rules. Moreover, when it becomes known that integrity plays no role and you will always choose the lesser of evils as between immediate consequences, you open yourself to blackmail by those who play dirty games. In the terms of our Central American example, how do you know that the officer will keep his word and release the other man? Do you really want the local army to believe that it can successfully entrap visitors into doing their dirty work? When it becomes known that people will always choose the lesser evil in any situation, a "Gresham's law" of bad moral choices may drive out the prospect of good ones. Once you allow departure from rules and integrity, are you not on a slippery slope to rationalizing anything?

It seems there is much to be said for considering both rules and the weighing of consequences in moral reasoning. When we turn away from abstract philosophers' arguments, we notice that many people make such complex eclectic moral judgments on a daily basis. In practice many people judge moral virtue in a balanced, nonabsolutist way that would have been familiar to Aristotle in ancient Greece, even though the particular virtues are somewhat different in modern society. They tend to weigh three dimensions of motives, means, and consequences when reaching ethical as well as legal judgments. For example, suppose a well-intentioned person is trying to bring your child home on time on an icy evening. She speeds, the car skids off the road, and your child is killed. Her motives were good, but the consequences horrible because of her inattention to means and facts. It is not murder, but it may be negligent homicide. Her good intent reduces the charge, but it does not exonerate her. Our moral judgment is one of degree, not a binary choice of completely wrong or completely right.

### Three-Dimensional Ethics

Motives, means, and consequences are all important. Careful appraisal of facts and weighing of uncertainties along all three dimensions are critical to good moral reasoning. Right versus wrong is often less difficult to handle than right versus right and degrees of wrong. We can be properly critical of "one-dimensional" moral reasoning, which ignores the complexity of many large moral issues that characterize international relations.

A good example of one-dimensional moral reasoning is the case of those who equated the American intervention in Grenada with the Soviet intervention in Afghanistan. In some of their motives—namely to maintain a sphere of influence—the actions of the two superpowers were similar. But the bloodiness of the means and the probable consequences (in terms of restoring local autonomy) were quite different. In a three-dimensional moral perspective, Grenada and Afghanistan were very different. Similarly, on a one-dimensional approach, the American intervention in the Dominican Republic in 1965 and the Soviet intervention in Czechoslovakia in 1968 had similarities in motivation, but the American action was partially redeemed by the eventual consequences of creating a more autonomous and democratic Dominican society, while Czechoslovakia has lost on both counts. On the other hand, good consequences alone are not sufficient to make an action good. If a murderer is trying to kill me and I am saved because a second murderer kills my would-be assailant first, the consequences are good, but the action is not. An invasion that has fortunate consequences is better than one with disastrous consequences, but a three-dimensional judgment might still judge it as a morally flawed action.

That was part of the problem with the American intervention in Vietnam. Norman Podhoretz has argued that our involvement was moral because we were trying to save the South Vietnamese from totalitarianism.[36] The people who led us were those who had learned from the Munich experience that totalitarian aggression must be resisted even if it is costly. But if American idealism was part of the cause of our role in the Vietnam War, that same idealism tended to blind our leaders to the facts of polycentric communism and local nationalism as alternative means to America's less idealistic end of perserving a balance of power in Asia. It also blinded them to the unintended consequences involved in a guerrilla war in an alien culture and the immoral effects that would follow from the disproportion between our goals and our means. On a three-dimensional judgment, our involvement in Vietnam was morally flawed.

Some three-dimensional judgments are easy, and some are extremely difficult. Consider the admittedly simple mix of cases in Table 1.

If we believe that good motives are a necessary though

TABLE 1
*Three Dimensions of Moral Judgment*

|  | MOTIVES | MEANS | CONSEQUENCES |
|---|---|---|---|
| *Case 1* | Good | Bad | Bad |
| *Case 2* | Good | Good | Good |
| *Case 3* | Good | Bad | Good |
| *Case 4* | Good | Good | Bad |
| *Case 5* | Bad | Bad | Good |
| *Case 6* | Bad | Good | Bad |
| *Case 7* | Bad | Bad | Bad |
| *Case 8* | Bad | Good | Good |

not sufficient condition for moral praise, then the last four cases are easy. The first two cases are also simple. It is the third and fourth cases that cause the greatest difficulty (though some might argue that in a nuclear world cases 5 and 8 are better than 4). In practice one would have to look at particular circumstances and weight consequences against other dimensions. Rather than decide *a priori* whether case 3 or 4 is more moral, we would do better to think of them as representing a continuum of hard choices trading off means and consequences. Moral virtue would consist in the care, the quality of moral reasoning, and the procedures that went into weighing such choices, rather than an arbitrary assignment into one or the other philosophical tradition. My argument is summarized in Figure 1. Moral praise would

FIGURE 1.
*Judging Moral Integrity*

Good and Bad Means ◄───────────────► Good and Bad Consequences

*Moral integrity is a disposition toward:*
1. Standards of clarity, logic and consistency
2. Impartiality (i.e., respect for the interests of others)
3. Initial presumption in favor of rules and rights
4. Procedures for protecting impartiality
5. Prudence in calculating consequences

be for the person or a society disposed toward careful moral reasoning that was respectful of all three dimensions.

Moral integrity would not mean locating a point of equal distance or equilibrium between means and consequences, but a disposition toward certain qualities in the moral reasoning used in making tradeoffs. The first two criteria in Figure 1 have already been discussed. The other three still need explanation.

If every rule can have its exception, how do we protect against too easy a collapse into consequentialism? Even those who base their ultimate arguments on consequences should beware of premature or shallow consequentialism. How does one introduce handholds or stopping points on the "slippery slope?" Two devices help. The first is always to start with a strong presumption in favor of rules and place a substantial burden of proof upon those who wish to turn too quickly to consequentialist arguments. That burden must include a test of proportionality, which weighs the consequences of departure from normal rules not only in the immediate case but also in terms of the probable long-run effects on the system of rules. For particularly heinous practices such as torture or nuclear war, the presumption may be near absolute, and the burden of proof may require proof "beyond reasonable doubt."

The other device (and the fourth criterion) is to develop procedures that protect the impartiality at the core of moral reasoning, which is so vulnerable in the transition from the rule-oriented to the weighing of consequences. Thought experiments or mental exercises that look at an action from the perspective of others is a good example. Such a mental reversal of position is what the golden rule urges upon us. In addition, developing ways to consult or inform third parties in order to protect against selfish assumptions is useful. Democratic procedures are not a guarantee of moral action, but they can often be of help. While secrecy is sometimes essential in strategic interactions, it has moral costs. The practice of consulting courts, Congressional committees, allies, and other countries can all serve as means to protect impartiality. In other words, while there is no perfect procedure for incorporating rules in a sophisticated consequentialist approach, this presumptivist and procedural approach is less self-serving and more likely to be impartial.

The fifth criterion is prudence in the calculation of consequences. It is impossible to know all the consequences of our actions. The more complex the situation, the more likely it is that the good consequences we intend may be swamped by unintended evil consequences. Prudence in the calculation of consequences is essential to protect against wishful thinking that can produce great evil. When an expected consequence depends upon a long chain of uncertain events, we must expect the unexpected. There must be a reasonable prospect of success before we use intended consequences to justify an action. As we shall see below, prudence in the estimation of consequences is particularly important when uncertainties and potential disaster are as large as they are on nuclear issues.

The disposition for careful balancing of moral dimensions according to these five criteria is not the only way to try to reconcile the different moral considerations. Many rule-oriented philosophers would allow considerations of consequences in cases of utter necessity or supreme emergency. For example, the Catholic Bishops of France appealed to an "ethic of distress," and the Catholic Bishops of West Germany discussed an "emergency set of ethics" in their justification of nuclear deterrence in 1983.[37] Michael Walzer argues that the human-rights-based principle of not killing civilians can give way to a principle of necessity in a "supreme emergency." In that approach the traditional rules that limited war no longer hold, and nuclear deterrence is justified as a means of coping with a permanent condition of supreme emergency.[38] But as others have observed, this is an awkward solution. "To see the entire postwar era as constituting this kind of moral emergency" creates a large gap in a rule-based theory.[39]

Sophisticated consequentialists can also find ways to reconcile the different moral dimensions. There is a difference between someone who tries to assess the consequences of each act alone and another who takes a broader framework. Sophisticated consequentialists will consider the wider and longer-term consequences of valuing both integrity of motives and rules that constrain means. They will also realize the critical role of rules in maintaining moral standards in complex institutions. A sophisticated consequentialist analysis would take the view of an "institutional utilitarian," asking the question, "If I override normal moral rules because it will lead to better consequences in this case, will I be damaging the institution by eroding moral rules in a manner which will lead to worse consequences in future cases?"

How do we reconcile rules and consideration of consequences in practice? One way is to treat rules as *prima facie* moral duties and to appeal to a consequentialist critical level of moral reasoning to judge competing moral claims. For example, in judging the moral acceptability of social institutions and policies (including nuclear deterrence), a broad consequentialist might demand that the benefit they produce be not only large but also not achievable by an alternative that would respect rules.[40] In addition, to protect against the basic difficulties of comparing different people's interests when making utilitarian calculations, a broad consequentialist would require very substantial majorities; otherwise he would base his decisions on rules and rights-based grounds.[41]

A consequentialist argument can also be provided for

giving some weight to motives as well as means. For example, William Safire argues that "the protection of acting in good faith, with no malicious intent, is what makes decision-making possible. It applies to all of us. . . . The doctor who undertakes a risky operation, the lawyer who gambles on an unorthodox defense to save his client, the businessman who bets the company on a new product."[42] While such an argument can be abused if good motives are treated as an automatic one-dimensional exculpation, it can be used by broad consequentialists as a grounds for including evaluation of motives in the overall judgment of an act.

Whether one accepts the broad consequentialist approach or chooses some other, more eclectic way to include and reconcile the three dimensions of complex moral issues,[43] there will often be a sense of uneasiness about the answers, not just because of the complexity of the problems "but simply that there is no satisfactory solution to these issues—at least none that appears to avoid in practice what most men would still regard as an intolerable sacrifice of value."[44] When value is sacrificed, there is often the problem of "dirty hands." Not all ethical decisions are pure ones. The absolutist may avoid the problem of dirty hands, but often at the cost of having no hands at all. Moral theory cannot be "rounded off and made complete and tidy." That is part of the modern human condition. But that does not exempt us from making difficult moral choices.[45]

# CHAPTER THREE

---

# Obligations to Foreigners

JUDGING MORAL reasoning is difficult in domestic set-
tings; it is more so in international relations. To whom
do we owe moral obligations, and to what extent? Most
U.S. nuclear strategists are "utilitarians with limits." They
think in consequentialist terms, but tend to consider only
the consequences for Americans (or their allies) rather
than the interests of Soviets, third countries, or humanity
at large.[46] On the other hand many activists and moralists
tend to be cosmopolitan in scope and refer to the interests
of humanity at large. If nuclear weapons could destroy
human life, they ask, how can we set national limits on
our thinking? They tend to neglect, or simply deplore,
the fact that the world is organized into nations and is
likely to continue in that form for quite some time. That

circumstance has always been difficult for those who wish to think carefully about moral issues in international relations.

There are four basic views about the moral obligations we owe to people who are not our fellow citizens. I have already argued that one view, that of the total skeptic who denies any duties beyond borders, rests on premises that do not stand up to careful scrutiny. When we turn to those who admit that obligations are owed to foreigners, we encounter three quite different schools of thought: the realist, the state moralist, and the cosmopolitan.[47] The realist and the state moralist tend to stress the value of order; the cosmopolitan values individual justice more highly.

## The Realist Approach

In contrast with the total skeptic, the realist accepts some moral obligations to foreigners, but only of a minimal sort related to the immoral consequences of disorder. The realist would argue that more extensive obligations exist only where there is community which defines and recognizes rights and duties. Such communities exist only in weak forms at the international level, and that sets strict bounds on international morality. Moreover, the world of sovereign states is a world of self-help without the moderating effects of a common executive, legislature, or judiciary. In such a domain chaos is the greatest danger. The range of moral choices is severely constricted, because the government that attempts to indulge a broad range of moral preferences may fail in its primary duty of preserving order. As Hans Morgenthau has written, "The state has no right to let its moral disapprobation . . . get in the way of successful political action, itself inspired by the moral principle of national survival . . . Realism, then, considers prudence . . . to be the supreme virtue in politics."[48]

This form of realism can be distinguished from the position of the total skeptic discussed earlier. The realist places the greatest stress on national survival and on the instrumental value of order. The most significant means of preserving international order is the balance of power. International politics is characterized by such inequality and by so little structure that on consequentialist grounds, maintaining the balance of power deserves a strong priority over both interstate and individual justice. Prudent pursuit of self-interest will at least produce order and avert disastrously immoral consequences, which the unbridled pursuit of justice might produce in a world of unequal states. This is accentuated by the existence of nuclear weapons. We owe foreigners the minimal obligation of order that avoids chaos leading to nuclear war. Henry Kissinger often replied to critics who accused him of an amoral policy because of his lack of zeal in pursuit of human rights, "Peace is the moral priority."

The realist has a strong argument in reminding us that one thing justice entails is a degree of safety for national populations, and that international moral crusades can lead to disorder, injustice, and evil consequences. But while it is true that the problem of order makes international politics less hospitable ground than domestic politics for moral arguments about justice, it does not follow that justice is totally excluded in the international realm. Order may be a necessary condition for justice, but it need not be treated as an absolute value. One can assign a high value to order and still admit degrees of tradeoff between order and justice. For example, imagine that a prominent Chinese visitor applied for political asylum in the United States, and the Chinese Government warned that failure to send her home would have serious consequences for Chinese–American relations. Granting asylum might then have some small effect on the balance of power in Asia and on the prospects for international order. If we treated order as an absolute value, we could send her back. But most Americans would prefer to exchange that small effect on order against the gain for individual rights. Not all cases are so easy, but the tradeoff is clear.

Both statesmen and citizens constantly weigh such tradeoffs in international affairs. As I said earlier, national survival may come first, but much of international politics does not concern survival. Choices among alternative courses of action must be made on many other issues, large and small, and those choices can be (and are) informed by many moral values. Moreover, while the balance of power is the primary means of preserving order in international politics, it is not the only means or a totally unambigous means. There are no strong institutions to enforce norms, but weak international institutions of law and diplomacy do supplement the balance of power in preserving some degree of order.

## The State Moralist Approach

The second approach stresses morality among states and the significance of state sovereignty and self-determination. One variant of the state moralist viewpoint stresses a just order among a society of states. For example, the philosopher John Rawls asks what rules the states would choose or would have chosen for just relations among themselves if they did not know in advance how strong or wealthy they would be.[49] The principles Rawls derives—self-determination, nonintervention, and obligation to keep treaties—are analogous to existing principles of international law.

Justice among states would not necessarily produce justice for individuals. Michael Walzer seeks to alleviate that moral problem by portraying the rights of states as a collective form of their citizens' individual rights to life and liberty. The nation-state may be seen as a pooled expression of individual rights. It represents "the rights of contemporary men and women to live as members of a historic community and to express their inherited culture through political forms worked out among themselves."[50] Thus there is a strong presumption against outside intervention.

That presumption, however, is not absolute. Foreigners have an obligation to refrain from intervention unless the lack of fit between a government and the community it represents is "radically apparent." Thus, for example, Walzer would allow intervention to prevent massacre and enslavement; to balance a prior intervention in a civil war; or to assist secession movements that have demonstrated their representative character. In such circumstances it would be contradictory to regard the state as representing the pooled rights of its individual citizens. This presumption and its exceptions are again analogous to many of the existing rules of international law. But it does not guarantee either realist order or cosmopolitan individual justice.

The virtue of Walzer's refined state moralist approach is that it bases the rights of states on the rights of individuals, while taking account of the reality of the way international politics is structured and conforming quite closely to existing principles of international law. A weakness in the approach is the ambiguity of the concept of self-determination.[51] What is the self that determines? How do we know when there is a "radical lack of fit" between a government and its people? Must an oppressed group fight and prevail to demonstrate its claim to speak as a people worthy of international recognition? If so, is not might making right?

31

As one of his critics asks, "In Walzer's world, are there not self-identified political, economic, ethnic, or religious groups (for example, capitalists, democrats, communists, Moslems, the desperately poor) who would favor foreign intervention over Walzer's brand of national autonomy (and individual rights) if it would advance the set of rights, values, or interests at the core of their understanding of justice? . . . Why should Walzer's individual right to national autonomy be more basic than other human rights, such as freedom from terror, torture, material deprivation, illiteracy, and suppressed speech . . .? Walzer's ideal is but one normative, philosophical conception among others, no more grounded and often less grounded in peoples' actual moral attitudes (and social identities) than other conceptions."[52] In short, the state moralist approach is particularly weak when it treats self-determination and national sovereignty as absolute principles that must come first in a rank ordering. In practice, peoples do want self-determination and autonomy, but they want other values as well. There is a constant problem of tradeoff and balancing competing moral claims between autonomy and other values.

### The Cosmopolitan Approach

The third approach stresses the common nature of humanity. States and boundaries exist, but their existence does not endow them with moral significance. Ought does not follow from is. David Luban has written, "The rights of security and subsistence . . . are necessary for the enjoyment of any other rights at all. No one can do without them. Basic rights, therefore, are universal. They are not respecters of political boundaries, and require a universalist politics to implement them; even when this means breaching the wall of state sovereignty."[53] Many citizens hold multiple loyalties to several communities at the same time. They may wish their governments to follow policies that give expression to the rights and duties engendered by other communities in addition to those structured at the national level.

While the cosmopolitan approach has the virtue of accepting transnational realities and avoids the sanctification of the nation-state, an unsophisticated cosmopolitanism also has serious drawbacks. First, if morality is about choice, then to underestimate the significance of states and boundaries is to fail to take into account the main features of the real setting in which choices must be made. To pursue individual justice at the cost of survival or to launch human rights crusades that cannot hope to be fulfilled, yet interfere with prudential concerns about order, may lead to immoral consequences. And if such actions, for example the promotion of human rights in Eastern Europe, were to lead to crises and an unintended nuclear war, the consequences might be the ultimate immorality. Applying ethics to foreign policy is more than merely constructing philosophical arguments; it must be relevant to the international domain in which moral choice is to be exercised.

The other problem with an unsophisticated cosmopolitan approach is ethical; it discards the moral dimension of national politics. As Stanley Hoffmann has written, "States may be no more than collections of individuals and borders may be mere facts. But a moral significance is attached to them."[54] People wish to live in historic communities and autonomously to express their own political choices. A pure cosmopolitan view that ignores those rights of self-determination fails to do justice to the difficult job of balancing rights in the international realm. One of the reasons that states have nuclear weapons is that peoples wish to defend their sovereign autonomy as independent moral communities at this stage in human history.

## A Cosmopolitan–Realist Synthesis

The argument among realists, cosmopolitans, and state moralists is over how to balance transnational and national values. Some of the main differences are summarized in Table 2. Different people may approach the balancing of such values in different ways, but the statesman's trusteeship role requires giving priority (though not exclusive attention) to interstate order and national interests. Sophisticated realists admit that justice affects the legitimacy of order, sophisticated cosmopolitans admit the political significance of boundaries, and sophisticated state moralists admit the possibility of duties beyond borders. As a result, the three positions in practice often tend to converge. Each has a part of the wisdom that must be considered in balancing competing moral claims in hard cases. The position I adopt could be called a "cosmopolitan–realist" approach. It accepts transnational obligations, but in a manner limited by the realities of the way the world is organized into states at this stage in history.[55]

Such a position is based on the acceptance of certain minimal obligations of common humanity. If one wishes to avoid controversial arguments about natural law, it is safer to ground arguments about rights and obligations in the existence (and nonexistence) of a sense of community. Where a sense of community exists, it is possible to define rights and obligations. There are multiple senses of community in world politics today, but those which transcend national boundaries in universal form are relatively weak. Nonetheless, at a minimum many people,

TABLE 2
*Four Views of International Ethics*

| | DUTIES | KEY VALUE | CENTRAL CONCEPT | INTERVENTION? |
|---|---|---|---|---|
| Skeptic | None | National interest | Raison *d'état* | To pursue advantage |
| Realist | Minimal | Order | Balance of power | To maintain balance |
| State moralist | Limited | Self-determination | Society of states | Nonintervention |
| Cosmopolitan | Large | Justice | Society of persons | To create justice |

including realists, acknowledge at least a weak form of community symbolized by the notion of "common humanity." Defining another being as part of the human community entails restraints on how we treat him or her. For example, we do not kill them for food or pleasure (as is done with animals by those who do not define community in terms of all creatures than can feel pain).[56] Nor does one allow them to starve if one is in a position to help. The outpouring of concern in response to pictures of starvation in Ethiopia is a case in point.

It is still debatable what rights and obligations follow from such a minimal form of community as the notion of "common humanity." Rights and their correlative obligations can involve modest claims (being left free from interference) or strong claims (being provided with something one does not have). A minimal sense of community may give rise only to limited rights and obligations. Walzer argues, for example, that "the idea of distributive justice presupposes a bounded world, a community within which distributions take place, a group of people committed to dividing, exchanging, and sharing, first of all among themselves. It is possible to imagine such a group extended to include the entire human race, but no such extension has yet been achieved. For the present, we live in smaller distributive communities."[57]

Some cosmopolitans would reply that the fact that the current sense of community is strongest at the border of the nation-state does not mean it is right to draw conclusions about limited obligations.[58] "Ought" does not follow from "is." The sense of community in world politics was more limited at times in the past and may be more expansive at times in the future. Rather than derive norms from current facts, one should derive them from an ideal sense of human community. That would be more consistent with the criterion of impartiality that is an essential characteristic of moral reasoning. If one were to engage in a mental experiment such as John Rawls suggested, and imagine what principles of justice all people (rather than states) might agree to if they were behind a "veil of ignorance" regarding their actual advantages, one would not allow enormous inequalities of wealth based on nationality. From that point of view, the burden of proof rests with those who depart from the cosmopolitan ideal. They must justify their preference to compatriots.

Although philosophers point out correctly that "ought" does not follow from "is," the facts about what exists severely shape and constrain normative judgments. If they are also right that "ought" implies "can," then "is" and "ought" are logically distinct but empirically intertwined. To apply Rawls's concepts of justice without regard to boundaries (which Rawls himself eschews) is a debatable procedure.[59]

Impartiality is not the same as egalitarianism. It does not mean "each one equals one." Impartiality means similar actions in similar situations, not identical actions. It prohibits moral justification based simply on the egoism of the actor.[60] It does not exclude considerations of the interests of the actor if similar interests would be allowed weight for other similar actors. It is not morally defensible to justify an action "because it is me." It may sometimes be right to justify an action "because the object is mine." For example, if you could save only one of two drowning children, one of whom is your own, you may be morally justified in saving your own. Your child has a right to expect such a duty from you on the basis of the roles of parent and child in the family community relationship. Of course, you would have to admit the same justification for any parent similarly placed, even if you were the absent parent of the child not saved.

Social roles create and carry rights and duties, whether the roles be those of family or of citizenship. But given

37

the existence of multiple levels of community and multiple roles, the preference for your own child or compatriot cannot morally be admitted as absolute. If you assumed the role of lifeguard and then noticed the drowning children, you would have an additional obligation to save the child where there was a higher probability of success. And at some point a consequentialist would object to a preference for saving your own child that did not consider the relative probabilities of success or the numbers of other children involved. In short, some preference for family or compatriots may be justified on grounds of impartiality, but it would not be possible to reconcile an absolute preference with that criterion.

What obligations do we owe to foreigners in the world as it now exists? Imagine a thought experiment in which you and other rational people were trying to answer that question before the deck of national cards was dealt and you did not know if you would be dealt a high (rich, strong nation) or low (poor, weak nation) card. Assume we (1) accept a sense of common humanity, but (2) have a preference for autonomous national community (or assume that it is the only feasible form of organization at this time), and (3) realize the dangers and difficulties of preserving order in a world organized into states with nuclear weapons. But (4) we lack a common ideal or vision of the good, and our concern for order must often have priority over justice. Even in this imperfect world, we would wish to set some limits on national exclusivity and preference for compatriots. If we knew we all had that dangerous preference for (or accepted the existence of) national communal identity at this point in history, but did not know whether we would belong to a rich, strong nation or a poor, weak one, what limitations would we want all nations to follow on preference to compatriots, and what obligations would we wish all nations to accept toward foreigners? The result might well be a sense of minimal obligations to persons outside one's nation.

First, when we recognize each other as part of common humanity despite national differences, we admit negative duties not to kill, enslave, or destroy the autonomy of other peoples as part of our definition of the term "humanity." While community generally implies reciprocal awareness of obligation, reciprocity may not be necessary to justify adherence to those duties. Even if another people lacks a sense of common humanity at this time, our definition of them as human and a thought experiment about impartiality would produce such restraints. We do unto others as we *would have* them do unto us. Even in wartime, a cosmopolitan–realist could accept the morality of limits on killing of innocent civilians (and other restraints of just war theory) despite the failure of the enemy to observe such restraints. Whether he would live up to this moral standard or succumb to psychological pressures to respond in kind is another question. But the moral standard is clear.

For example, even if fundamentalist Moslems in Iran in 1979 had regarded Americans as subhuman infidels who could be killed, since we accepted Iranians as fellow humans, we admitted certain restraints on our actions toward them. Popular songs in America may have had lyrics about "nuking Iran" during the hostage crisis, but there was no widely accepted or serious argument for such a disproportionate retaliation.

The negative duties include an obligation not to intervene in other states, because such intervention destroys the autonomy of the common life of another community. We could disregard prohibitions on intervention in situations where genocide, enslavement, or egregious deprivation of human rights made a mockery of the *prima facie* assumption that the autonomous political process in that state represented the pooled rights of the individual citizens. Deciding whether such conditions exist will be debated in light of the facts in particular instances. But while agreeing that such conditions would release us from the negative duty of nonintervention, it would not necessarily create a positive duty of intervention, particularly if such intervention would be costly or dangerous in a nuclear world. Positive duties exist, but they are limited.

A second minimal obligation to foreigners (and limit on our preference for compatriots) is the generally accepted consequentialist principle of taking responsibility for the consequences of our actions. It may be that "the great majority of actions that occur within the boundaries of a nation-state are not either the direct or indirect results of the actions of those who are outside its national borders."[61] But as studies of interdependence show, even outside the nuclear issue many actions and conditions that affect the prospects for justice within a nation are affected by actions of others outside the nation.[62] Such interdependence is a matter of degree, and effects are often difficult to ascertain. Moreover, there are thorny questions about time and a moral "statute of limitations."[63] If we believe in individual moral choice, children should not be held guilty of the sins of their forefathers. Nonetheless, according to the principle of responsibility for the consequences of our actions, we have obligations to foreigners in some proportion to the strength of the effects we are able to ascertain as ours. The use of nuclear weapons would certainly create such effects!

Those are not the only obligations to foreigners that one could derive from such a thought experiment. One could imagine minimal positive duties of samaritanism and charity in situations where foreigners could be made better off without making compatriots significantly worse off.[64] But it seems unlikely that the participants in our thought experiment would agree on extensive positive obligations if they threatened to lead to disorder.

Realist concerns about the unintended consequences of disorder imply limits on more extensive obligation, but they do not prohibit actions beyond the call of duty where they can be prudently reconciled with interstate order. And they do not prevent individuals from taking personal charitable actions that their state might have to eschew for reasons of interstate order. Nor do they prevent citizens from encouraging the evolution of a stronger sense of community beyond the nation-state for the future. As I shall argue in the concluding chapter, a realistic thought experiment about the limited obligations beyond borders in the imperfect world we now inhabit does not preclude more idealistic thought experiments about worlds that might evolve in the future. But it does not permit one to act as if those worlds were about to exist now. A thought experiment about our obligations to foreigners does not accept the blind national limits that characterize most strategic thinkers, but it also sets limits on the degree of cosmopolitan obligation so long as the world remains organized as it is now. On the one hand, if we accept minimal obligations of common humanity, the possession of nuclear weapons creates certain moral obligations to foreigners. On the other hand, it does no good to rail against the sovereign state as the cause of our nuclear predicament and to treat its near-term abolition as a realistic solution. At this stage of history, moral reasoning about nuclear weapons must involve a more subtle balancing of obligations to compatriots and obligations to foreigners.

# CHAPTER FOUR

---

# Nuclear Ends and Means

Nuclear weapons raise thorny problems about both ends and means. When we speak about ends, we are referring to the goals we seek; the consequences we both want and expect to follow from our actions. Our ends affect the motives for our actions.[65] Is there any end or goal that could justify such an action as killing large numbers of people in war? Self-defense is the common answer. Absolute pacifists would simply answer "no," but they are often accused of incoherence by consequentialists. If you refuse to defend yourself from imminent harm or death, you will have no more choices. As a position of absolute personal integrity, pacifism is not incoherent. Some people would rather die than do something they regard as wrong. But to the extent that people take responsibility for all the consequences of refusing to defend themselves and their societies in their moral reasoning, absolute pacifism seems less coherent and less compelling.[66] Moreover, as Augustine pointed out many centuries ago, it is hard to argue that a pacifist leader who refuses self-defense has a right to make that decision for other people.

While there is a long tradition of Christian pacifism that leads some, like Archbishop Raymond Hunthausen, to believe that "one obvious meaning of the cross is unilateral disarmament,"[67] the dominant Western tradition has been the just war doctrine. The just war tradition represents a middle path between pacifism and militarism that has evolved over hundreds of years of Western history. In the words of the American Catholic Bishops' pastoral letter, "governments threatened by armed unjust aggression *must* defend their people. This includes defense by armed force if necessary as a last resort."[68] Or as Pope John Paul II put it, "those who possess a sense of reality and a love for true freedom and dignity of individuals and of nations are thus convinced of the right to defend oneself against an unjust aggressor."[69]

Some people are wary about the idea of "just war." For example, some philosophers fear that the concept could legitimize ideological crusades, and history makes them leery of all crusades. As one German philosopher put it to me, "if you had our history and our location, you too would worry about any doctrine with war in the title."[70] At first glance the concept might seem to legitimize "wars of national liberation" (which are just wars in the Soviet lexicon) or a fundamentalist Islamic "jihad." But the words "just war" give a misleading impression of a doctrine that actually establishes a very restrictive set of conditions which limit the use of force. In current usage, crusades are ruled out, and the doctrine might better be understood as a "just defense" doctrine.

Even if properly understood as tight restrictions on the use of force, some writers believe that nuclear weapons have exploded the just war doctrine. They argue that no end can justify nuclear war, and that nuclear weapons are immoral means. In that view, the two main aspects of just war doctrine, *jus ad bellum* or the right to go to war, and *jus in bello* or limits on the means used in war, are both called in question by the unprecedented destructiveness of nuclear weapons. I shall look first at the arguments about ends, and later in the chapter at the arguments about means.

Motives and Ends

There has been much less discussion about ends than about means in the current nuclear debate. *Jus ad bellum* has received far less attention. According to James Johnson, "in recent debate moral analysis has simply ignored the two requirements that have traditionally engaged most moral thought: just cause and right intent."[71] In the eyes of some critics, the American Bishops "only belatedly and inadequately mention the ends of nuclear deterrence and defense. Their discussion of war is almost exclusively an exercise in the morality of means, divorced from the ends that are the sole warrant for even considering the means in question. To consider the morality of means without considering the morality of ends is a 'stunted' approach."[72] Such criticisms often come from those who feel that the prevailing moral arguments pay too little heed to the nature of the Soviet threat. In terms of the criteria discussed in Chapter 2, they have a valid point that ends and means are best discussed in relation to each other.

Is there any end that could justify a nuclear war that threatens the survival of the species? Is not all-out nuclear war just as self-contradictory in the real world as pacifism is accused of being? Some people argue that "we are required to undergo gross injustice that will break many souls sooner than ourselves be the authors of mass murder."[73] Still others say that "when a person makes survival the highest value, he has declared that there is nothing he will not betray. But for a civilization to sacrifice itself makes no sense since there are not survivors to give meaning to the sacrifical act. In that case, survival may be worth betrayal." Is it possible to avoid the "moral calamity of a policy like unilateral disarmament that forces us to choose between being dead or red (while increasing the chances of both)"?[74]

How one judges the issue of ends can be affected by how one poses the questions. If one asks "what is worth a billion lives (or the survival of the species)," it is natural to resist contemplating a positive answer. But suppose one asks, "is it possible to imagine any threat to our civilization and values that would justify raising the threat to a billion lives from one in ten thousand to one in a thousand for a specific period?" Then there are several plausible answers, including a democratic way of life and cherished freedoms that give meaning to life beyond mere survival. When we pursue several values simultaneously, we face the fact that they often conflict and that we face difficult tradeoffs. If we make one value absolute in priority, we are likely to get that value and little else. Survival is a necessary condition for the enjoyment of other values, but that does not make it sufficient. Logical priority does not make it an absolute value. Few people act as though survival were an absolute value in their personal lives, or they would never enter an automobile. We can give survival of the species a very high priority without giving it the paralyzing status of an absolute value. Some degree of risk is unavoidable if individuals or societies are to avoid paralysis and enhance the quality of life beyond mere survival. The degree of that risk is a justifiable topic of both prudential and moral reasoning.

Even within a common religious tradition, people of goodwill can differ in their judgments. In 1983 the American, German, and French Catholic bishops focused on the twin goals of protection of justice and prevention of war. All three found just cause for the possession of nuclear weapons. "Americans need have no illusions about the Soviet system of repression and the lack of respect in that system for human rights or about Soviet covert operations and pro-revolutionary activities." And the French bishops argued that "it would be unfair to simply state and accept the conflict of ideologies while closing one's eyes to the domineering and aggressive character of Marxism-Leninism."[75] But the American bishops dwelt far less on the issue of protecting justice and conditioned their acceptance of deterrence on its being used only for prevention of nuclear war. The French and German bishops dwelt more extensively on the geopolitical threat to liberty and justice and were willing to see nuclear deterrence used to prevent conventional as well as nuclear war. Different situations and different perceptions can contribute to different weightings of goals and acceptable risks, even when values are shared.

Historically, a variety of causes were accepted in the just war doctrine. Just wars could include punishment, the restitution of something wrongly taken, and religion. For example, medieval crusades to defend the faith were called just wars at that time. But "international law in the twentieth century has gradually reduced the justifications of war to one: defense."[76] And modern theologians tend to portray just war doctrine as intermediate between pacifism and crusades.[77] Self-defense is a broadly accepted right, but a right of self-defense is not unlimited. The use of force must be a last resort in the face of imminent harm, and then carried out with means that are limited and proportionate to the values being defended. (Whether such limits can be observed in nuclear war is discussed below).

Both crusades to promote a faith and nuclear strikes out of vengeance fail the test of proportion between defense of critical values and risks to survival. "A moral community does not attack out of vengefulness or self-righteousness conjoined with a just cause."[78] Suppose the Soviet Union were to launch a full-scale nuclear attack that destroyed our cities and killed a large part of our people? The idea of justice as retribution would suggest a response in kind. But if such retribution would lead to destruction of the species (or civilization in the Northern Hemisphere), there would be an inadequate proportionality between that value and the threat to survival. It might make sense to deter and punish Soviet leaders by targeting nuclear weapons on them (as I shall discuss below), but not target weapons in a manner that would destroy all men, women, and children.

Unfortunately, even when we limit just cause to self-defense it is a rather elastic concept. There are few weapons that are purely offensive or defensive. And in a bipolar nuclear balance of power, there is an astonishingly broad range of actions in remote areas that can be related to self-defense. The justification of deterrence as self-defense cannot be separated from the broader issues of the justification of a nation's foreign policy. Sometimes political perceptions of a delicate nuclear balance of power are invoked to justify far-fetched or marginal foreign policy goals. But to resort to nuclear threats in order to protect low stakes is a morally and politically nasty bluff.[79] Fortunately, prudence reinforces virtue in helping to limit such threats, since in deterrence a particular move is likely to succeed only if it is sufficiently proportionate to crucial values that it will appear credible.[80]

Faced with those ambiguities, some people tend to base their view of just cause solely on the virtue and motives of their country. For example, Caspar Weinberger told Oxford students in 1984 that American foreign policy was justified because it "can be changed by the voters." To which a student replied, "if [one is] beaten and tortured by those regimes [which American supports], is it a more moral act because Congress approves of it instead of some general?" On the other hand, when Weinberger's opponent in the Oxford debate, E. P. Thompson, asserted that "the United States and the Soviet Union speak exactly the same language," he was guilty of placing too little rather than too much emphasis on ends.[81] American freedoms are not sufficient, but neither are they irrelevant to just cause.[81] Such arguments marked the European debate over deployment of NATO's intermediate range nuclear weapons in 1983. Opponents argued for European neutralism: Both superpowers and their missiles were equally bad. Proponents argued that the missiles could not be divorced from the ends they served, and that the Americans showed far more intent to preserve political liberties and human rights than the Soviets did.

The best protection against spurious moral reasoning based on motives is to engage in a thought experiment that encourages a degree of impartiality. If one played the thought experiment discussed in the previous chapter and did not know one's nationality, what goals would be worth defending, and at what levels of nuclear risk? Other nations would probably accept some nuclear risks imposed upon them for the sake of self-defense and minimal order, but they would be unwilling to see those risks seriously raised by an expansive policy or global crusade for American-defined values.

For example, many Europeans (and Americans) felt that when President Reagan defined support for Nicaraguan counterrevolutionaries as "self-defense" in his 1985 State of the Union address, he was stretching the meaning of the concept.[82] When Americans formulate their ends very loosely, foreigners fear that the scale of means (and risks) will also become expansive. Self-defense may serve as a just cause for the possession of nuclear weapons, but a simple thought experiment about the views of others shows that it must be narrowly defined and embedded in a prudent approach to foreign policy. Ends and means may be considered separately, but the hard questions about nuclear self-defense will be questions of proportionality. Different situations will lead people to different judgments. Efforts to insert impartiality into moral reasoning can help to provide a common point of discourse. Further on I shall return to the limits on self-defense.

## Means

Philosophers in the Kantian tradition tend to focus on the problem of means, and the principle that one must not do evil as a means of doing good. Consequentialists are not convinced by that principle. For example, Robert W. Tucker argues that even if there are no absolute limits on means, there are still differences among means and questions of how *much* evil one may do that any good may come. "Instead of restraining our behavior, the belief that evil may never be done may only strain our ingenuity."[83] Some arguments about deterrence do indeed have that quality. For example, those who object on moral grounds to the targeting of civilians but do not object to our targeting of sixty military sites inside Moscow because we are not deliberately targeting civilians are simply straining their ingenuity!

Some take an absolutist position on means. The Catholic

theologian Germain Grisez, for example, does not doubt that if we dismantle our strategic deterrent, "the U.S.S.R. would reduce us and other Western nations to puppet status," but "the issue is not our readiness to suffer evil, but rather our willingness to do it. The murderous intent of the deterrent is a moral evil. . . . Better anything than mortal sin."[84] Others avoid the problems of absolutism and appeal to conceptions of human rights as well as personal intuitions about integrity to justify following their principle in all but rare catastrophic situations. We do not accept slavery, colonialism, or vicarious punishment of the innocent even if they could be proved to have beneficial consequences. Similarly we should not accept deterrence based on nuclear weapons that threaten disproportionate devastation and harm to innocent people.

Proportionality and discrimination between combatants and noncombatants are the two key criteria by which just war theory judges the legitimacy of means, or *jus in bello*. Do nuclear weapons fall into those two categories of means that are prohibited because they are indiscriminate or disproportionate? Some say a positive answer is inherent in the mass destructiveness of nuclear technology. For example, Jonathan Schell argues that "immorality is inherent in the very possession of tens of thousands of nuclear weapons, whatever the doctrine. There is no conceivable way that these can be used without mass slaughter on an incalculable scale, and no theoretical sophistry can eliminate this basic fact."[85] But the critics tend too quickly to assume that any nuclear war is the same as all-out nuclear war and that any failure of deterrence must be a total failure. That need not be the case. They try to win their point without having to look carefully at all the relevant facts.[86]

Catastrophe is not necessarily inherent in nuclear technology. It is quite possible to think of uses of nuclear weapons that do not violate the *jus in bello* criteria. As for proportionality of destruction, nuclear warheads such as the "neutron bomb" can be coupled with precision guided delivery systems and airburst above tanks so that they would do less damage than some conventional shells used in the two world wars and deposit very little radioactive fallout. And nuclear weapons used at sea on naval warfare targets could absolutely observe the principle of discrimination between combatants and noncombatants.

The critical question, of course, is not technological impossibility but whether those limits can be maintained once the key political taboo has been broken. Some critics assert categorically that "any serious threat or use of nuclear weapons is immoral. The most critical reason for this is their intrinsic uncontrollability."[87] Or as another philosopher has put it, the critical moral feature of nuclear weapons that frustrates their limitation by the just war criteria is their "technological recalcitrance" to the intentions of their users.[88]

Certainly technological slipperiness is good cause for great prudence. The Catholic bishops were right to express skepticism about claims that a nuclear war could be limited, and proponents of carefully controlled nuclear-war-fighting strategies often make fatuous assumptions about rationality and control, which make their arguments totally unrealistic. But it does not follow politically or technically that the chance of limitation is zero. Technically, a key factor will be the ability to maintain command, control, and communication. That will be affected by the targets chosen and the size and number of nuclear explosions. It is plausible to imagine control and communication surviving dozens or even hundreds of explosions; it is implausible to imagine them surviving thousands of explosions (or even more limited attacks) that include command centers as targets.[89]

Politically, it is not implausible to believe that the first response of leaders to a limited nuclear use may be to keep it limited and settle the conflict within which it occurred.[95] That does not mean nuclear use would be prudent or wise. The opposite effect of anger, confusion, and escalation could also occur. The prospects of maintaining control can never be certain. But likewise the prospects of limitation after nuclear use are not zero. To believe otherwise would imply that the prospects of limitation are equal in all cases. That flies in the face of what we know about the technology and about strategic doctrines. It is implausible to assume that escalation to all-out nuclear war would be equally likely in the case of an American nuclear attack on a Soviet submarine or a Soviet brigade invading Iran, and the destruction of Moscow.[91] The risks and uncertainties may make it wise for a leader to treat the prospects of control as extremely low, but not as impossible.

Why do such distinctions matter if prudence counsels skepticism about more than minimal claims regarding controllability? The answer lies in the problem of use in relation to deterrence, which is the key dilemma of nuclear weapons. Given the existence of some risk of unintended horrible consequences, one should wish to avoid any use of nuclear weapons. But if there is absolutely no possibility of the use of nuclear weapons, or if that is believed to be the case, they will have no deterrent effects. Thus deterrence depends on some prospect of use, and use involves some risk that just war limits will not be observed.

What types of use and what degree of risk constitute acceptable means is a matter on which reasonable people can differ. The American bishops, for example, rule out first use or any use against civilian targets. But they do not exclude second use against military targets. While they caution against assuming such use can surely be controlled, the prospect of second-strike counterforce use provides the "centimeter of ambiguity" that they believe is needed to sustain the role of the deterrent as a means of protecting just ends. Thus they avoid impaling themselves on one horn of the "usability dilemma," though their critics feel they crept too close to it. Their critics argue that such a restricted possibility of use is simply "a fig leaf covering a small area of nakedness of the 'use, never' position" and is too little to deter potential aggression.[92] Indeed, some Europeans believe that opposition to first use weakens the nuclear deterrence of conventional war and that escalation from a conventional war is the most likely occasion for nuclear war.

Whatever their conclusions about nuclear use, not all who reason within the just war tradition come to the same conclusions about deterrence as about use. In the view of the French bishops, use would be immoral, but *threat does not constitute use.* [93] That distinction suggests solving the moral dilemma by threatening without really intending to carry out the threat. Bluff deterrence might be a nice solution to the "usability dilemma" if it could work. And a president probably cannot fully know his intent to use nuclear weapons, or not until a particular occasion arises. Robert McNamara has suggested that there may have been an element of bluff in the American deterrent during the period he served as Secretary of Defense in the 1960s.[94]

But a thoroughgoing bluff deterrent would be virtually impossible to implement. Deterrence requires a complex bureaucratic machinery to which it would be impossible to convey the intention of bluffing without disclosing the bluff to the adversary and thus defeating the strategy. Even if the bluff is kept solely within the president's mind, the machinery must be prepared and exercised, and provisions must be made for successors in command if the president is killed. By succession or by accident, the machinery might lead to nuclear war regardless of the president's intentions. A private hypothesis of bluff or uncertainty about use might salve his conscience, but it could never serve as a national strategy that would solve the moral dilemmas of his subordinates or of the citizenry.

For some philosophers, not even a successful bluff would be acceptable if one accepts the principle that "it is wrong to threaten what it would be wrong to do." But that principle is not self-evident, and it seems too restrictive, for one can imagine spurious threats that have the effect of saving lives without any intent of doing evil. For example, the principle would prevent you from threatening harm to a man who is paralyzed by fear in order to make him leave a burning building.

Philosophers might make the principle more plausible by stating that "it is wrong to intend to do what one knows it is wrong to do."[95] But even in this refined formulation, the principle is too simple an approach to strategic action to be able to capture the situation of deterrence.[96] Deterrence is a game with more than one player. In strategic interactions, outcomes depend on two sets of intentions, not just one's own. One could argue that one's own intention is to preserve peace by a threat of nuclear retaliation, and the outbreak of nuclear war would depend on the opponents' actions, not one's own intentions. Consequentialists argue that in strategic interaction, "intentions may have autonomous effects that are independent of the intended acts actually being performed," and "these effects must be incorporated into any adequate moral analysis of it."[97] Strict Kantians then must respond that even conditional or indirect intentions to do evil cannot be justified regardless of their consequences.

Even if one grants that there must be a genuine, if conditional, intent of nuclear use if deterrence is to be credible, their simple moral argument is not fully compelling. As Bernard Williams says, it implies that having a deterrent strategy and having a nuclear war are morally

54

equivalent. "Purely in the abstract, the argument does not follow."[98]

A specific example concerns the just war principle of discrimination and its injunction not to do harm to innocent persons. Germain Grisez considers "the U.S. deterrent absolutely immoral. This judgment is made not on pacifist grounds and not by arguing better-Red-than-dead, but by refusing either to pretend that one can will effective deterrence without choosing the death of civilians or to try to rationalize that wickedness."[99]

Concern for the innocent is not only a long-standing principle that has been used to save lives in war, but it goes to the heart of the just war doctrine's willingness to allow any killing. Killing is justified only to save life from an imminent threat. Those who present no threat of harm cannot be attacked under the rationale of self-defense. "Innocent" does not refer to the moral condition of the victim, it refers to whether they pose a threat of harm that justifies killing them. For example, once a soldier throws down his gun, it is not permissible to shoot him even though he is in uniform.

Large-scale modern warfare had blurred the line between combatants and noncombatants even before the advent of nuclear weapons. The woman working in a munitions factory may be no more "innocent" of threat than a young man conscripted into the infantry. But it is hard to imagine a threat posed by young children that justifies self-defense by threat of nuclear retaliation. One might argue that they benefit from family and social life in the threatening nation, which implies tacit consent in that nation's policies. But that stretches the difficult notion of tacit consent beyond any reasonable interpretation. And it does nothing to address the dangers that nuclear deterrence may impose on innocent people in third countries, which are not allied to and do not threaten the superpowers.

One way philosophers have tried to escape the problems associated with innocent life is through the doctrine called "double effect." By this doctrine, if an action has both good and evil consequences, it may be excused so long as the evil was not intended as an end or as a means. One may not choose evil as a means to do good, but evil that is the indirect or unintended consequence of good means can be morally acceptable. For example, if a soldier were being fired on from a farmhouse, and after destroying the house by mortar fire he found that he had killed four soldiers and a child, the death of the child could be excused under double effect. But the doctrine would not justify his seizing the child and threatening to harm it as a means to force the soldiers to surrender.

Both the American Catholic bishops and the Protestant theologian Paul Ramsey have used the doctrine of double effect to justify deterrence so long as civilians are not deliberately targeted. There must be no intent to kill innocents, though some innocents may be killed. Many philosophers are uneasy with the doctrine.[100] If intent includes not merely the motives that impel one to act, but also a willingness to produce clearly foreseeable consequences, the excuse of double effect seems morally arbitrary in some cases. Destroying a city because it was near a nuclear target is not like the unforeseen killing of the child in the farmhouse.

Moreover, double effect can lead to easy abuse in the nuclear age. At one time, Strategic Air Command planners chose military targets for their "bonus effects" on civilians, although the official doctrine was counterforce.[101] And when sixty military targets are inside the Moscow city limits, what moral difference does it make if we say we do not target civilians and any collateral damage is an indirect and unintended consequence?

A better way to approach the problem of respecting innocent life is to treat it as a relative principle. A Catholic student of the just war tradition, William O'Brien, suggests treating the principle of discrimination in relative rather than absolute terms. He argues that "literal application of the principle of discrimination is incompatible with nuclear war, as it is with virtually any kind of modern war. . . . Discrimination is not an ironclad principle. It is a relative prescription that enjoins us to concentrate our attacks on military objectives and to minimize our destruction of non-combatants. . . . The standard of judging the sufficiency of the effort to minimize civilian damage is one of proportionality."[102] Not everyone agrees. Some philosophers complain that O'Brien's approach reduces the principle of discrimination solely to that of proportionality, which is more elastic and less satisfying from a rule-oriented perspective.[103] But if one wishes to apply the principle of discrimination in just war theory to nuclear deterrence, O'Brien's suggestion makes sense. And the principle is worth maintaining both because of its close relation to just cause and because of the intrinsic value of minimizing (if not completely avoiding) harm to innocents.

Some consequentialists throw up their hands at the amount of ink that Kantian philosophers spill in intramural quarrels about means. Russell Hardin argues that the focus on the individual agent's actions misses the institutional factors that are the important features of deterrence. Robert Tucker argues "the prospects for fulfilling the requirements of *bellum justum* can be no more than rhetorical."[104] It is a focus on means gone wild. But even if it cannot be treated in absolute terms, the principles of just war doctrine represent a long moral tradition of wrestling with issues of means that should not be discarded lightly.

Some of the arguments about just war principles resemble medieval philosophers debating angels on the head of a pin (or missile), but the arguments appeal to historically important rules that prevent leaders from submitting too easily to overly elastic utilitarian standards. The just war doctrine is worth preserving in modified form as a set of restrictive principles for a just deterrent. I shall use it to suggest specific maxims in Chapter 7. More generally, as Gregory Kavka has argued, extreme utilitarians and extreme Kantians will not be satisfied by a modified approach, but "for a system of morality to reflect our firmest and deepest convictions adequately, it must represent a middle ground between these extremes by seeking to accommodate the valid insights of both act-oriented and agent-oriented perspectives."[105] A modified just war doctrine can help us to develop such an approach to nuclear ends and means.

# CHAPTER FIVE

# Consequences and Risks

MOTIVES AND MEANS are only two dimensions of moral reasoning. Consequences are the third, and many philosophers as well as practical politicians believe that the consequences are the most important criterion by which the morality of nuclear policies should be judged. When the potential consequences are so enormous, "otherwise honorable concerns with perfection, virtue, rights, and the doctrine of double effect simply give way. The difference between letting humanity or some large part of it be immolated and causing it to be immolated is a moral difference that pales into insignificance."[106]

It is sometimes assumed that consequentialists always favor nuclear deterrence because deterrence has worked, and that rule-oriented philosophers tend to oppose deterrence because it involves immoral means. A stylized argument has a theologian saying that deterrence is wrong because it threatens harm to innocents and a consequentialist replying that deterrence is good because it has worked for forty years and thus has saved many innocent lives. But the correlation is far from perfect. As we saw in the last chapter, some rule-oriented philosophers believe nuclear deterrence can be justified within the just war tradition. And some consequentialists oppose deterrence because human fallibility and "Murphy's Law" make failure seem inevitable, and the effects of failure may be catastrophic.[107]

Recent discussions about the prospect of "nuclear winter" have reinforced this critical strand of consequentialist thought. Nuclear winter is the theory that the smoke created by burning cities in a full-scale nuclear attack would block the sunlight and cause a drop in temperature that might eventually destroy life on this planet. There is a great deal of scientific uncertainty about the theory of nuclear winter, and the threshold at which it might occur is far from clear. A few hundred weapons dropped on cities in summer might have such an effect; a few thousand striking missile silos in winter might not. While the uncertainties about nuclear winter are great, its possibility cannot be ruled out, and it has helped to reinforce a recent spate of antinuclear consequentialism, some of it quite apocalyptic in tone. The writings of Jonathan Schell are a good example.

In this chapter I shall look at the arguments put forward by both the pro-deterrence and the antinuclear consequentialists. Both revolve around the argument that deterrence seems to have worked for forty years. In a nutshell, the pro-deterrence consequentialists make too much and the antinuclear consequentialists make too little of the absence of any war between the superpowers for the past four decades. The simple form of the pro-nuclear argument begs the difficult causal question of ascertaining what would have been the case in the absence of nuclear deterrence, and more important, it also begs the question of why the future should resemble the past. For those who believe that catastrophic failure is inevitable, it is no answer to say simply that deterrence has worked in the past. In addition to potential catastrophic failure, the antinuclear consequentialists charge that if deterrence is working, nuclear weapons are having serious adverse effects on our daily social and political life.[108]

At the same time, the argument that deterrence has worked is not without merit. Four decades without war among the great powers is a remarkable period of peace in modern Western history. (The record is forty-three years between 1871 and 1914).[109] And while they are not the only causes of peace—memories of world war and the fact that neither of the big victors of 1945 was deeply dissatisfied also played a role—there is good reason to believe that nuclear weapons contributed to the prudence that kept leaders out of war over the past forty years.[110] Nuclear weapons may have had something like a crystal ball effect. Imagine what would have happened if the statesmen who led the world into this century's first great conflagration in 1914 had possessed a crystal ball showing them the world of 1918. The leaders of 1914 expected a short, sharp war, followed by business as usual. One suspects that if the German Kaiser, the Russian Tsar and Austrian Emperor had seen a picture of 1918—with their thrones vacant, their empires destroyed—they would have drawn back from the brink of war that summer. Today modern leaders know the horrible devastation that would result from any nuclear war.

But that is only partial comfort, because crystal balls can be shattered by accident or miscalculation. Even if nuclear deterrence has lasted for nearly four decades, it is difficult to believe that it will last forever. Some dire fears, such as C. P. Snow's 1960 prediction that nuclear war within a decade was a mathematical certainty, have proved wrong. It has been argued that Snow's exaggeration will be justified if nuclear war takes place in the next hundred years.[111]

Pro-deterrence consequentialists argue that such a view is too pessimistic, because the difference between a decade and a century can make all the difference in the world when it comes to deciding what can be done to avoid nuclear war. In the view of pro-deterrence consequentialists, careful analysis is required to try to clarify the possible causation and evaluate the risks of nuclear war. Some acceptance of risk is necessary if we are to realize values beyond mere survival. By the evidence of their daily behavior, few people make survival an absolute value in their personal lives. The same can be said for societies. Consequentialist *moral* reasoning is concerned about relating risks and values; about the reduction of disproportionate risk and the fair distribution of risk.

### Antinuclear Consequentialism and Inevitability

Antinuclear consequentialists often object to the whole approach to deterrence in terms of probability. For example, the sociologist Todd Gitlin argues that "since deterrence works only if it works forever, it is an all-or-nothing proposition, so applying the language of probability to it is misleading."[112] But his argument is not compelling. Gitlin seems to assume that failures of deterrence or inevitable accidents must lead to all-out nuclear war, but that is far from self-evident. Indeed, a case can be made that an accident or partial failure of deterrence may be the prelude to substantial changes that reduce risks or reliance on nuclear deterrence in the long term. But even if he were right about catastrophe, it is odd for Gitlin to discount "microscopic probability" by asking, "Do we feel secure playing Russian roulette if the revolver has a hundred chambers?" Perhaps not, but if we had to play, I doubt that we really would not care if a revolver had one hundred chambers rather than six! And if he readmits probabilistic reasoning, then it can also be applied to the question of relative risks between unilateral disarmament ("refusing to play") and trying to increase the number of chambers in a world where the game of nuclear deterrence already exists.

Jonathan Schell provides another version of the same argument. "The very existence of uncertainty about whether or not a holocaust would extinguish our species should lead us to treat the issue morally and politically *as though* it were a certainty. . . . Morally they are the same."[113] If we treat this language as poetic license exhorting prudence, it may have great meaning, but if taken literally it is a nonsequitur. By a verbal sleight-of-hand, he moves from uncertainty to certainty, and then having banished the key question of probability and risk, he slips on to a dubious moral judgment of equivalence. Neither his logic nor the facts about "nuclear winter" can support that conclusion. Estimates of the effects of "nuclear winter" include an enormous range of uncertainty, and extinction of the species is far from a certainty.[114] Moreover, nuclear weapons are not the only threat to the species. It may turn out that biological warfare, which produces and distributes new lethal molecules, may become cheaper and more dangerous than nuclear weapons in the next century. And nuclear weapons might play some role in deterring such toxic warfare.

Suppose, however, we accept it as a certainty that an all-out nuclear war would lead to extinction of our species. Would it follow that "we have no right to gamble because if we lose, the game will be over, and neither we nor anyone else will ever get another chance."[115] It does not follow from the fact that extinction is an unlimited consequence that even a tiny probability is intolerable and that our generation has no right to take risks.

The issue raises interesting problems about obligations among generations. What obligations do we owe to future generations whose very existence will be affected by our risks? A crude utilitarian calculation would suggest that since the pleasures of future generations may last infinitely (or until the sun burns out), no risk that we take to assure certain values for our generation can compare with almost infinite value in the future. Thus we have no right to take such risks. In effect, such an approach would establish a dictatorship of future generations over the present one. The only permissible role for our generation would be biological procreation. If we care about other values in addition to survival, this crude utilitarian approach produces intolerable consequences for the current generation.

Moreover, utility is too crude a concept to support such a calculation. We have little idea of what utility will mean to generations very distant from ours. We think we know something about our children, and perhaps our grandchildren, but what will people value 8,000 years from now? If we do not know, then there is the ironic prospect that something we deny ourselves now for the sake of a future generation may be of little value to them. A more defensible approach to the issue of justice among generations is the principle of equal access. Each generation should have roughly equal access to important values. We must admit that we shall not be certain of the detailed preferences of increasingly distant generations, but we can assume that they will wish equal chances of survival. On the other hand, there is no reason to assume that they would want survival as a sole value any more than the current generation does. On the contrary, if they would wish equal access to other values that give meaning to life, we could infer that they might wish us to take some risks of species extinction in order to provide them equal access to those values. If we have benefited from "life, liberty and the pursuit of happiness," why should we assume that the next generation would want only life?

The equal access approach assumes that each generation would wish to make the tradeoffs for themselves. The current generation cannot avoid imposing some risks upon the future. As Derek Parfit argues, the risk does not do injustice to identifiable persons, since they do not yet exist. Later the harm may become real. Nonetheless, if the risks are kept low and values are successfully preserved, the gamble benefits a next generation, who then make their own decisions about risks and benefits to be passed on to further generations. Keeping risks to the survival of the species at a low level is essential to a sense of proportionality. Survival is not an absolute value, but it is important because it is a necessary condition for the enjoyment of other values. The loss of political values may (or may not) be reversed with the passage of time. The extinction of the species would be irreversible.[116] Thus proportionality requires that we rate survival very highly, but it does not require the absence of all risk.

Proportionality in risks is easier to judge if we think in terms of passing the future to our children and letting them do the same for their children rather than trying to aggregate the interests of centuries of unknown (and perhaps nonexistant) people at this time. While the contemplation of species extinction—or what Schell calls "double death"—may reduce the meaning of life to some people in the current generation, that is a value to be judged against others in assessing the risks that are worth running for this generation.[117] It is not a cause of injustice to a future generation.

Antinuclear consequentialism often uses the concept

of probability to bolster its case that deterrence will inevitably fail. With a series of trials over time, even a low probability approaches a certainty. The probability of at least one occurrence in a series of n trials is equal to $1 - (1 - P)n$.[118] Thus if we flip a coin once, the chance of getting tails at least once is 50 percent. If we flip a true coin ten times, the chances of seeing tails at least once is 99.9 percent. Using this approach to the distribution of probabilities, Bradford Lyttle concludes that the average daily chances of the unauthorized launch of a nuclear missile over the past thirty years was probably no higher than .0006 percent. But Lyttle goes on to argue that even if the daily chances of one missile being launched are only one in a hundred million, the probability of such an event passes 50 percent within forty years.[119] Or Douglas Lackey points out that a one in a hundred chance of nuclear war in the next forty years becomes a 99 percent probability after eight thousand years.

While such calculations can be useful numerical reminders, we cannot conclude very much about deterrence from them. They assume that probabilities are constant, that events are independent of each other, and that the world of the future will be much like ours. But the metaphor of a flipped coin is misleading. Human interactions are more like loaded dice. The odds change, and the outcome of one set of events may greatly change the odds for the next event. In fact, frightening events like the Berlin or Cuban missile crises may drive the odds of war down in their immediate aftermath. A diplomatic rapprochement such as occurred between the United States and China in the 1970s can have a similar effect. New technologies may cut either way. It seems unlikely that the odds of nuclear war faced by future generations will be the same as ours.

Alternatively, some antinuclear consequentialists who believe in the inevitability of nuclear war make the opposite assumption, that the odds are not independent but are driven in a perverse direction by the interaction of events.[120] But unless this is an example of the "gambler's fallacy" that a series of events is evidence for an increased probability of more such events ("That guy's on a roll."), there must be a specified causal model to make such an argument plausible. Such causal models can be evaluated in the light of evidence. Often, like the proposition that arms races inevitably lead to war or that we have been piling up more and more destructive capacity, the models turn out to be weak on the facts.[121] Historically, many arms races have ended without leading to war.[122] Statisticians give the example of a baseball pitcher. "At the start of a season a pitcher may not be in good physical condition, and his probability of throwing a strike may be low. . . . Later a small injury may plague him. For this pitcher, one may well believe that the probability of throwing a strike will wax and wane with time. But simple forms of the law of averages are not readily applicable to such a complicated process."[123]

The likelihood of nuclear war rests on both independent and interdependent probabilities that relate to different aspects of the process by which war might occur. Purely accidental war might be conceived of in terms of independent probabilities, but if the numbers are low enough, they may not matter. If we are speaking of eight thousand years, for example, humankind may have concerns other than nuclear war. And colonies in space will probably exist. The longer the period, the greater the chance that other things will have changed and that an accident will have a different meaning in that changed context. In short, even when we have independent probabilities (as we might approximate in the case of pure accidents) the significance of the events they lead to must be seen in the context of the interdependent probabilities involved in human history. As Paul Schroeder has put it, "Murphy's Law does not apply to history." That is not an argument for optimism, but it is "an argument *against* a certain extremely popular kind of crippling fatalism."[124] Consequentialists like Gitlin and Schell who use simple probabilistic arguments to argue that abolition of nuclear weapons is the only policy that is both moral and realistic have built their case on shaky grounds.

A more sophisticated antinuclear argument against probabilistic reasoning is presented by the British philosopher Robert Goodin. He argues that all "notions of probability and likelihood are simply inappropriate, "because we just do not know enough about the shape of the underlying distribution to justify employing any of the standard techniques for estimating probabilities. . . . Thirty-five years is just too short a run." What is more, says Goodin, we lack well-validated scientific theories about nuclear war from which we might derive probabilities, and psychological evidence shows that individuals are poor probability assessors. Thus nuclear deterrence is reckless "playing the odds without knowing the odds." Instead of probabilities we should use a "possibilistic" logic with only three crude categories—certain, impossible, and possible—to assess consequences.[125]

Goodin's points are a useful caution against spurious precision in estimates of probabilities that strategists often make. He warns us against the efforts to "fine-tune" deterrence and to justify huge expenditures on new weapons systems by tiny increments in deterrence. But Goodin's arguments are less convincing as a reason to lump all probabilities into one crude category of "equally possible." When we deal with rare events like nuclear wars, where there can be no series of tests, probability is bound to be subjective. But that does not make the concept useless. The chance that a certain football team will win a championship may be one in a million, yet it is possible to find a bookie who will take a bet. "In such a circumstance, that is exactly what is meant by probability, the odds that a rational and informed person will set on the occurrence of the event. . . . It is, in fact, an informed guess."[126]

Our knowledge about nuclear deterrence is limited, but it is not totally nonexistent. There is evidence from forty years of experience. Theory is weak in international relations, but it is not totally absent. And the evidence from psychological experiments that Goodin quotes show that subjective estimates of probability are often wrong, but they also show that estimates by specialists tend to be more reliable than estimates by the general public. Moreover, the public tends to overestimate unfamiliar risks (while underestimating familiar ones).[127] And as Russell Hardin has argued, when differences in expected outcomes are large (as they are with nuclear issues) "we need not have a very accurate sense of the probabilities we face—very crude estimates may suffice."[128] Our ignorance is not so great that it justifies Goodin's conclusion that nuclear disarmament is a "moral certainty."

Douglas Lackey, an antideterrence utilitarian philosopher, would accept the proposition that rough estimates of probability are sufficient to make judgments about deterrence. He uses such calculations to argue that utilitarians should advocate a policy of unilateral disarmament.[129] Unlike Schell and others, Lackey admits that such a policy might significantly increase the likelihood of limited Soviet nuclear strikes against the United States. After all, the only time nuclear weapons have been used in war was against a country that could not retaliate in kind (and by a country we think is more constrained by moral considerations than we believe the Soviet Union would be!). Not only might the Soviets use nuclear weapons to destroy our conventional forces, but they might destroy a city to enforce blackmail. And they would have an enormous incentive to destroy any facilities that would enable us to recover our nuclear capability. Since occupation of the United States would be enormously difficult, such a nuclear strategy might look very cost-effective.

Nonetheless, Lackey calculates that in terms of expected lives lost, the increased probability of suffering a limited nuclear attack under unilateral disarmament will be multiplied by a much lower number of lives lost than if we pursue strategies of nuclear deterrence that could lead to all-out nuclear war. A utilitarian who focuses on expected lives lost might ask what risk of nuclear war would be worth tolerating if it meant avoiding a near certainty of conventional world war every twenty years. Lackey concludes that unilateral disarmament is the most moral policy—even if nuclear weapons help deter conventional wars, which, in the case of World War II, may have cost some 50 million lives.

But why should a consequentialist focus solely on expected lives lost except as an escape from the problem of balancing incommensurable values? We are back to the problem of trading off survival against other values. What Lackey has done is simply given us a more elegant version of "Red versus dead," dressed up in the language of probability. The structure of the problem is not changed. What Lackey's exercise does is remind us that probabilities matter when tradeoffs are being weighed. If the probabilities of nuclear war look high enough, many people, for example in Europe, might reasonably choose Soviet occupation over total devastation. But if the risks are low enough, they reach other conclusions. The success of deterrence policies in democracies depends on public perceptions of probabilities related to several values, not just expected lives lost.

## *Pro-Nuclear Deterrence Consequentialism and Risk*

Pro-deterrence consequentialists argue that it is essential to pay attention to the relative risks of deterrence and the alternatives to it. As Bernard Williams has said, "the world would be safer if there were not any nuclear weapons . . . but it does not follow that everyone ought to get rid of nuclear weapons. . . . The morality of deterrence is legitimately one in which you think principally about steps which make it less likely that the weapons get used. . . . The moral approach in my mind cannot avoid complex arguments about what the world is actually like."[130] Some estimate of the level and direction of change in the probability of nuclear war under various alternatives is fundamental to this approach to consequentialist moral reasoning. One might pursue risky policies like unilateral disarmament or early abolition if the probabilities of all-out war are extremely high. But if such probabilities are not high, such risky policies would not be justified. Of course, even a low-probability event could happen tomorrow. To return to our earlier example, one can be killed on the first shot in Russian roulette, but the number of chambers makes a big difference.

What are the risks of nuclear war? That is a critical question, but unfortunately there is no fully satisfactory answer. In recent years public opinion polls have shown anything from one-third to one-half of the American public saying they expect nuclear war with the Soviet Union within the next decade. In contrast, my informal polling of a few score nuclear specialists over the past few years has produced a modal answer of expectations of nuclear war between the United States and the Soviet Union in the next decade of about one chance in one hundred. The range of answers among experts, however, stretched from one in five to one in a million. For comparison, the known probability of being killed in an automobile accident in the same period is about one chance in five hundred.

The point is not to prove that the public or the experts are wrong but merely to remind us of the range of opinions in any estimates of the risks of nuclear war. Despite our inability to be precise, we are bound to make crude estimates of whether the probability is high or low. It seems more plausible to assume that the probabilities are relatively low than that they are very high. There are both rational and accidental models of the causes of war, and both models seem to have low probabilities in the nuclear age. As McGeorge Bundy has written, "in light of the certain prospect of retaliation, there has been literally no chance at all that any sane political authority, in either the United States or the Soviet Union, would consciously choose to start a nuclear war."[131] As for pure accident, the prospects seem low. In Michael Howard's words, it is hard to find any historical evidence of a purely accidental war.[132] And technical progress over the past two decades has reduced the prospects of purely accidental onset of nuclear war.[133] In other words, there are good theoretical reasons to accept the experience of the last forty years as evidence that the unknown probability is unlikely to be at the high end of the distribution of probabilities. We cannot say that the annual average probability was not higher than one in forty, but we can say on the basis of the distribution of probabilities that it was unlikely to be much higher than that. It is possible, but not likely, that the average probability of nuclear war has been high. Of course, average probabilities can be misleading if trends or unexpected events cause changes, but they provide a rough starting point or baseline. We should make neither too much nor too little of the forty years that we have experienced.

Unfortunately, it is also difficult to be precise about trends in probability over time. For example, is the current situation more or less risky than the period of previous concern, 1958–62? Those who argue that nuclear risk was higher in 1962 point to technical improvements such as electronic combination locks on weapons; improved command, control, and communication; national technical means of verification; and such political factors as U.S. and Soviet experience in managing crises. Those who argue that the risk is higher in the current period point to the loss of U.S. nuclear superiority; the greater Soviet capability to support forces in third world areas; the deployment of vulnerable weapons and support systems that place a premium on preemption; doctrinal stress on protracted war-fighting; and the deterioration of political dialogue.

In examining such arguments, it is interesting to note the mixture of technical and political factors, and the mixture of arguments resting upon assumptions about rational actors and those stressing nonrational and accidental factors. Policy responses vary accordingly.

Within a rational actor framework, the insanity of large-scale nuclear war (i.e., the extreme disproportion between political ends sought and the consequences of the military means used) suggests that nuclear war is very unlikely. At the same time political conflict will occur, and we shall need to deter a variety of Soviet actions. Hawks argue that the danger of nuclear war comes from a Soviet miscalculation of the credibility or capability of our commitment. They recall the appeasement of Hitler in the 1930s; Munich is their favored historical example. The way to reduce risk and to enhance deterrence is to add nuclear capability so that it is clear we cannot be beaten at the end of a multiple-move game (or escalation ladder).

The rational actor model can also lead to other conclusions. Many Doves tend to see provocation as a greater danger than temptation for the onset of war, and point out that the Hawks' unrestricted armament could create such provocation. While the historical evidence that arms races lead to war is far more shaky than prevalent views admit, there are historical cases where a country has intended to deter another but has succeeded only in provoking it to initiate war. For example, American efforts to deter Japan from advancing into Southeast Asia in 1941 instead provoked the attack on Pearl Harbor.[134] The proper response in such cases is reassurance, and Hawks and Doves differ on the relative dangers of provocation versus appeasement and the proper balance to be struck between deterrence and reassurance.[135]

A third view of the probability of the onset of nuclear war might be called the Owl's view. It rests on a nonrational model that includes such factors as psychic stress, misperception, bureaucratic pathologies, and accidents. "Owls" worry about loss of control more than wrong intention: 1914 and the onset of World War I is their dominant historical example. And while pure accidents under normal conditions are extremely unlikely, the probability of accident and inadvertent initiation can rise dramatically in a situation of nuclear crisis. At such times the safety catches that make accidents unlikely are deliberately released, psychic stress is increased, and there is little time to correct mistakes. And such crises also present the rare situation where, on an expected value definition of rationality, it might make sense to initiate war. If one were absolutely convinced that the other side were about to launch a nuclear attack, it would be better to go first rather than go second.

In short, the interaction of nonrational and rational factors in crisis situations that might involve escalation and preemption seem the most likely causes of dramatic rises in the probability of nuclear war. Following Williams's dictum that consequentialist moral reasoning should focus on "steps that make it less likely that the weapons got used," more attention should be focused on preventing and managing situations where there are sudden rises in probabilities. While all three models of the onset of nuclear war capture important aspects of causation, many strategists have focused too exclusively on policies of deterrence and reassurance related to intentions, and focus too little attention on situations where intentions may be irrelevant or so constrained that rational choice becomes virtually meaningless.

Should such crises arise, we may be glad to have rules and personal moral restraints play some role, and glad also to have antinuclear consequentialist critics like Goodin warn us against spurious precision in subjective estimates of probability. For example, it is reported that President Kennedy may have believed the chances of some type of war were between one in three and one in two at the depth of the Cuban Missile Crisis in 1962. Given the American strategic advantage at that time and the prospects of limiting damage to the United States by preemptively striking Soviet nuclear forces, some strategists might conclude that Kennedy must not have believed those probabilities would lead to a nuclear exchange, or else he acted irrationally in not launching the first strike that an expected value calculation would suggest. But Kennedy may have felt that such odds were still too low to justify a preemptive strike because he still had a significant chance of avoiding any nuclear war.

One might go further and ask whether expected value rationality makes sense, where choices are so constrained, stakes so high and uncertainty so great. It might make more sense to speak of nonrational situations that require suspension of efforts at precise calculation. If efforts to assess probabilities involve large uncertainties, and there is no clearly dominant choice in terms of consequences, the best moral reasoning may rely on Kantian rules or revert to the crude "possibilist" consequentialism that Goodin suggests. It is interesting to observe that whatever the truth about his real estimate of probabilities, they never reached the point of near-certainty that nuclear war was imminent, and Kennedy was reportedly influenced by moral intuitions against first strikes.[136]

Others have criticized Kennedy not for his choices in the depth of the crisis but for allowing it to develop in the first place. Was it worth raising the risks of nuclear war so high just to remove Soviet missiles from Cuba? Reasonable people may arrive at different answers, depending on their different attitudes toward the risk of great losses. But any answer must include an estimate of relative risks in light of the possible alternatives. Had Kennedy allowed Khrushchev to deceive us in Cuba, might Khrushchev subsequently have miscalculated American reactions somewhere else like Berlin, and would that not have increased the probability of a nuclear war? All risks must be weighed in the light of alternative outcomes.

Douglas Lackey complains that the public takes risk too lightly when disastrous outcomes do not transpire. "The risk of an evil is itself an evil," and it is fallacious to think that if worse does not come to worst, no harm was done. On expected value calculations, the harm done by risk is equal to the probability of the bad event times its disutility. "It is irrational not to take expected deaths seriously," yet people find real deaths less tolerable than expected deaths.[137] Faced with a situation like the Central American firing squad described in Chapter 2, many people would choose an act that would create a 10 percent chance of death for two hundred people rather than kill one (selected by lot) in order to save 199—even though each of the two hundred at risk might plead for the second choice. Lackey regards that choice as wrong, because he does not include considerations about personal integrity (such as those discussed in Chapter 2). But even in consequentialist terms, one might ask whether Lackey should not also consider the quality of a risk such as that involved in nuclear deterrence. Some risks make people more uncomfortable than others—for moral or other reasons—even though their expected value is the same.[138]

## *Other Consequences and Competing Moral Claims*

Qualititative dimensions of the risks of nuclear deterrence are among the costs that are alleged by another set of antinuclear consequentialists. They allege that even if nuclear deterrence works, it creates unacceptable psychological pain, corrupts the democratic political process, and diverts resources from better purposes.[139]

The issue of psychological "scarring" is an empirical question that cannot be solved *a priori*. Michael Walzer is correct to argue that it is misleading to liken deterrence to placing babies on the bumpers of automobiles as a means of encouraging safer driving. Such analogies blur the important difference that such a situation would be a much nastier experience for the babies than living under nuclear deterrence is for most citizens.[140] Perhaps all citizens are nuclear hostages today, but they are able to carry out their daily lives in a manner completely different from hostages held by hijackers or in an embassy held by terrorists. To equate nuclear terror with hijacking misses important differences in consequences.

Nonetheless, there is psychic pain associated with terror of nuclear weapons, and that is a cost to be charged against the benefits attributed to deterrence. Some children experience such fear more intensely than others, but that raises interesting questions about causation and appropriate responses. Robert Coles suggests that nuclear fear seems to be transmitted to children in large part by liberal middle-class parents.[141] If the probabilities of nuclear war are lower than such parents (or others) have suggested to their children, is the psychic pain to be attributed to deterrence or to the alarmist statements? Sometimes antinuclear activists justify exaggeration as necessary to awaken the American people. The hands on the doomsday clock are portrayed as approaching midnight. But is exaggeration morally justified if it causes unnecessary psychic pain to children and others?

The political effects of nuclear weapons are also a serious consideration. American values have been affected. Our central government is larger, and the executive branch plays a larger role in foreign affairs. Interaction between strategic adversaries involves secrecy, and secrecy is difficult to reconcile with democracy. Many of those changes began before 1945, but enormous life-and-death decisions are nevertheless delegated to the president or his successors, and the circumstances may not permit congressional involvement. Knowledge of the details of nuclear targeting plans tends to be restricted to the military, and there have been cases in the past where a significant gap has existed between military plans and what elected officials thought to be policy.[142]

Nonetheless, it is a serious exaggeration to say that "nuclearism has caused a cultural, as well as a political and constitutional, breakdown."[143] As Walzer points out, citizens don't understand the economy yet feel themselves competent to judge its management by its effects.[144] The relevant effects in the management of the nuclear issue is the perception of relative risks of war and risks to national values, and the public votes and acts accordingly. Indeed, whatever one's view of the Nuclear Freeze movement, its effect on the 1982 Congressional elections helped to push the Reagan Administration toward a somewhat more centrist nuclear position. Public opinion polls suggest that in general American deterrence policies are not far out of line with popular preferences, and when the public feels that an administration seems not to be protecting it adequately against nuclear war on the one side, and the Soviet Union on the other, it makes its views felt. That does not mean that the public is always adequately informed about details or that citizens can participate fully in the formulation of nuclear policies, but it does mean we should not exaggerate the political costs charged to the account of deterrence.

A more serious cost is the diversion of resources from competing moral claims to national defense. With one person in seven below the official poverty threshold in the United States, and even lower standards of living in the rest of the world, the competing claims on scarce resources are a weighty moral consideration. Large expenditures on exotic military systems that are justified by minute marginal effects on deterrence often are not adequately held to a standard of proportionality. At the same time the costs of sovereignty are high, and the consequences of misjudgment of what deters can be grievous. Most important, however, is the fact that nuclear weapons are only a small part (approximately 15 percent) of the defense budget, and efforts to reduce reliance on nuclear weapons are likely to increase rather than decrease the defense budget. Ironically, "nuclearism" may be better for competing moral claims on resources. But increasing reliance on the relative role of nuclear weapons in deterrence would be too risky to be a wise policy choice.

To summarize, the antinuclear consequentialist critics fail in the overall thrust of their argument. The argument that "deterrence has worked" is a powerful moral argument, but it is only the beginning, not the conclusion of adequate moral reasoning about nuclear weapons. It is a mistake to make either too much or too little of the fact that nuclear deterrence seems to have worked for forty years. Nuclear deterrence is not good *per se;* it is an instrument that must be judged in the light of alternatives. It does not follow that all deterrent policies in all situations can be justified. And while their general arguments fail, the antinuclear consequentialists succeed in their arguments to the extent that they caution us against spurious precision in the calculation of risk and failure to acknowledge the costs of competing moral claims.

Good consequentialist moral reasoning must rest on careful causal assessments of the relative risks of different deterrence policies and the alternatives to them, including an awareness of the broad bands of uncertainty that will necessarily be involved. It must relate risks to values for this and future generations, reduce disproportionate risk, and address the fair distribution of risk. And given the uncertainties in such calculations, consequentialists should welcome an element of rule-oriented thinking in constructing moral maxims for dealing with our nuclear dilemma. I turn next to the distribution of risks and the development of such rules of conduct.

# CHAPTER SIX

---

# Third Countries:
# Innocent and Not So

THIRD COUNTRIES raise special problems in nuclear ethics. In concluding that one can justify a deterrence policy with proportionate risks on consequentialist grounds, I have focused primarily on the U.S.–Soviet relationship. One can argue that each side gains certain benefits in self-defense in return for the risks they incur. But what right have we to impose risks on third countries? As Javier Perez de Cuellar asked the United Nations in 1984, "I see the delegations of 159 member nations . . . And all of them—all of us—live under the nuclear threat. As Secretary-General of this organization, with no allegiance except to the common interest, I feel the question may justifiably be put to the leading nuclear-weapon powers: by what right do they decide the fate of humanity?"[145]

Or in the words of the former Chancellor of West Germany, Willy Brandt, "it is unacceptable and terrifying" that people outside the United States and the Soviet Union must "depend for their right to live on a small group of people in one or two capitals."[146]

In the case of nations that are allied with us, we can argue that those countries consent to such risks. While there are differences about acceptable degrees and burdens of risk between for example, the United States and West Germany, the alliance rests upon consent. But that is not true of nonallies. Some philosophers argue that the essence of injustice consists in attempts to seize benefits while passing burdens on to others. As one philosopher put it, "The American people and Soviet people (it seems) are seizing the benefits of deterrence while passing the risks to third parties. Thus nuclear weapons systems certainly do inflict an injustice upon *them*."[147]

## Obligations to Those Who Pose No Threat

One might try to escape that conclusion by basing one's arguments for deterrence on amoral grounds of necessity or by invoking the doctrine of double effect, but as we saw earlier neither is a very satisfactory moral answer to the nuclear dilemma. Consider the scenario of nuclear winter. When it was first raised, some strategists argued that it made little difference to their thinking, since they had already long assumed that both the United States and the Soviet Union would lose their societies in an all-out nuclear war. But it is doubtful that Brazil, Nigeria, and Indonesia, for example, expected to lose theirs as well.

If, as argued above, we owe at least some obligation to foreigners as a condition of common humanity, then the narrow, nationally bounded calculation of some strategists is not morally acceptable. And to try to incorporate risks of nuclear winter under the category of unintended side effects that are excused by the principle of double effect is rather ludicrous. Douglas Lackey has drawn the analogy of a neighbor who is tired of being robbed and puts a huge stock of dynamite in his cellar and surrounds the house with elaborate tripwires and large signs warning intruders of the dangers of the system. If his house blows up, yours will too. He says, "I don't want to blow you up, but I have a right to keep out burglars and there's only a small chance that the dynamite will blow." Lackey doubts "that you would find the fact that your neighbor regretfully installed the dynamite a sufficient excuse for putting you at risk."[148]

Lackey's metaphor is constructed to suggest disproportionate risk. If one changed the metaphor to devices on the perimeter of a housing estate with warning signs, armed guards, and many barriers, the risk might not seem so high. Still, if we were discussing whether to install such a device, Lackey may be right. At the very least, if we consented at all, we would want some compensation for our risk (e.g. it should protect my house too) and some means of assuring that the risks are kept modest. Let us engage in a thought experiment to check the impartiality of our reasoning. Suppose it were 1939 and representatives of states were debating whether the United States should invent the bomb, but were doing so behind a veil of ignorance regarding their nationality. They would know that there was only a 6 percent chance that they would be Americans (i.e., our share of world population). They might say that according to the right of self-defense all should get the bomb or none should. But if they knew more about the real historical context and believed that an aggressive leader like Hitler might get the bomb, many might approve Roosevelt's decision that the United States make a nuclear weapon because the others would benefit through the maintenance of their independence.

Whatever the justice or injustice of the original American decision to invent the bomb, nuclear knowledge now exists, and moral reasoning must start from that fact. Some rule-oriented philosophers admit as much. For instance, even if it would be "absolutely immoral to *initiate* a policy of nuclear deterrence where none existed . . . it is not immoral to *continue* a policy of deterrence under certain conditions once such a policy exists. . . . In such cases, the absolutist Pauline principle that evil should not be done that good may come of it is impossible to apply. . . . In these cases the best that can be done is to choose the lesser of two evils."[149] It is as though one moved into a neighborhood where the dynamite in the basement system was already installed in several houses, and where there were substantial dangers of trying to dismantle them in the short term. What then would you demand of your neighbor? You might demand the right to install a similar system, but if you thought that would have the consequences of increasing your risks more than your benefits, you would not. You might, however, argue from the moral principle of accepting responsibility for the consequences of one's actions that your neighbor had an obligation to subject his device to the following conditions: (1) Use it only for defense against burglars (i.e., not to threaten neighbors). (2) Install devices to reduce the risks of accidents. (3) Provide some compensatory benefits (such as warning burglars that your house is protected too). (4) Take steps to dismantle the system at a time in the future when the incidence of burglary declines and relatively safe means of dismantlement are found.

An analogous process might occur if individuals in the world today were to assume a veil of ignorance about their nationality and then ask if the current unequal possession of nuclear devices could be morally justified. Since there would be about a 60 percent chance that they would belong to the population of one of the seven states (United States, Soviet Union, Britain, France, China, India, Israel) that may now possess nuclear weapons, they might well answer yes out of self-interest. But that is not a moral argument, and not all citizens of those countries agree with nuclear deterrence. If the individuals knew nothing of historical context and probable consequences, they might reason from the principle of an equal right to self-defense among sovereign states, that all or none should have nuclear weapons. But if they were informed that in the current historical context the efforts to get to either of those two conditions might significantly increase the risks of nuclear war, it is quite plausible that they would accept the unequal possession of nuclear weapons if certain conditions were met. Such conditions would be based on the principles of impartiality and accepting responsibility for the consequences of one's actions and might include that (1) the purposes of nuclear weapons be limited to self-defense (no imperial aggrandizement); (2) the weapons are treated with a special care that reduces risk of use; (3) there is some compensation in terms of preserving both independence and other values inherent in the order that the nuclear balance of power creates; and (4) steps be taken to reduce the risks (particularly to third parties) of reliance on nuclear deterrence, including dismantlement when conditions permit it to be done safely. In other words, the moral justification of the uneven possession of nuclear weapons depends on the existence of limits on ends and means, as well as continued attention to the relative risks created by deterrence and its alternatives. Those conditions are elaborated further in the next chapter.

## *Third Countries and Proliferation*

It is interesting to note that there exists in the current world a rough approximation of the imaginary compact we constructed in our thought experiment. Notwithstanding the equal right of self-defense among sovereign states, the great majority of nations (130) have adhered to the Non-Proliferation Treaty (NPT), which establishes two categories of states: five recognized as nuclear weapons states and the rest promising not to follow suit. It is also worth noting that the treaty involves two articles that imply compensation and risk reduction: Article 4, which provides for assistance in the development of peaceful uses of nuclear energy, and Article 6, which requires the nuclear weapons states to take steps toward disarmament. In addition, in various United Nations forums the superpowers have made limited pledges not to use nuclear weapons to threaten nonweapons states (unless the latter are acting in accord with another weapons state.) And it is interesting to note that in the war that followed Argentina's attack on the British position in the Falkland (Malvinas) Islands, Britain made no nuclear threats against nonnuclear Argentina. Nor did the United States seriously consider the use of nuclear weapons to avert its defeat by nonnuclear North Vietnam. In fact, there seem to have been only a handful of occasions where the United States seriously considered the use of nuclear weapons as a means to a political objective (rather than merely fearing that a crisis might lead to nuclear war), and those occurred in the 1950s before deterrence became effectively bilateral.[150] The taboo against the use of nuclear weapons seems to be firmly established among the states that currently possess them.

Of course, the current non-proliferation regime of treaties, norms, and taboos rests not solely on moral considerations but primarily on self-interest and prudence. Even though the superpowers have done an inadequate job of carrying out Article 6 of the NPT, a number of states continue to adhere to the treaty because they believe their security would be diminished if more states (particularly their regional rivals) obtained nuclear weapons. The treaty helps them reduce fears of cheating by their neighbors, because it provides for international inspections by the International Atomic Energy Agency to assure that peaceful nuclear programs are not being diverted to weapons purposes. And many countries realize that the existence of robust deterrence postures by the two superpowers is one of the reasons why some thirty states that by now could have chosen to make nuclear weapons have not done so. The credibility of the nuclear umbrella that the superpowers extend to their allies is a principal reason why proliferation has not spread nearly as fast as was expected two decades ago when President Kennedy predicted some twenty-five weapons states by now.[151]

A small number of important nuclear threshold countries (such as India, Pakistan, Argentina, Brazil, Israel, and South Africa) have refused to adhere to the NPT. They charge that it is discriminatory and hypocritical for the superpowers to maintain weapons that are denied to other states, and it is wrong for the superpowers to try to stop others from gaining access to nuclear weapons. Some observers argue that just as the existence of nuclear weapons has produced prudence that has stabilized the U.S.–Soviet relationship (the "crystal ball effect"), so would the spread of nuclear weapons to other countries similarly produce stability among regional rivals.[152] In this view, a world of nuclear porcupines would act like a world of well-behaved rabbits.

Other things being equal, that argument might have some merit. But other things are not equal, and the inequality has nothing to do with racial stereotypes or charges of irrationality in leaders of less developed countries. The problem with the argument is that it rests on a simplistic rational model of deterrence and ignores the "Owls'" concerns about inadvertent sources of war. The key difference is in the degree of risk that the "crystal ball" will be shattered. That risk exists in the U.S.–Soviet relationship, but I argued in the previous chapter that the risk has been relatively low. It is likely to be considerably higher in most regional situations because of the different experience, political conditions, and technical capabilities in the command and control of nuclear weapons, particularly in the early stages of a nuclear program, when new weapons are tempting and vulnerable targets for preemptive attack.

Looking at statistics of civil wars and overthrown governments; procedures for civil control of the military; and technical capabilities such as electronic safety locks and secure communication from the center to the field, it is plausible to believe that the risks of nuclear weapons' use among new proliferators is considerably higher than the risks of their use in the U.S.–Soviet relationship. The logic of the argument is the same in both cases. The superpowers also face risks that the crystal ball will shatter and should also take steps to reduce those risks. Nuclear weapons are dangerous for all people in the long run, but they are even worse in the short term for some people who lack certain command and control procedures and where there is a high frequency of conflict domestically and with neighboring countries. Nor can that problem be remedied by transfer of control technologies. Such transfers often would not be credible, and there is the risk that they would encourage further proliferation. Non-proliferation is not an inconsistent or hypocritical policy if it is based on impartial and realistic estimation of relative risks.

If a nonweapons state nonetheless decides to accept large risks, is that not purely its choice? Are other states justified in trying to prevent it from acquiring weapons? If the weapons were used on their own soil and posed a risk only to their own people, there might be no problem. But others are justified if the choice to obtain nuclear

88

weapons will impose significant new risks on third parties. There are several ways in which such consequences might arise. If new proliferators are more likely to use weapons—even if inadvertently—the breaking of a forty-year taboo and the danger that others might be drawn into the nuclear conflict increases risks.[153] And if the new proliferators have inadequacies in their procedures for control of weapons or of weapons-usable materials, such as plutonium or highly enriched uranium, the prospects that terrorists will gain access to nuclear devices also increases risks for everybody. And finally there is the simple but plausible proposition that the more nuclear weapons spread, the greater the prospects for eventual inadvertent or accidental use, the more difficult it will be to manage nuclear crises when many players are involved, and the greater the difficulty in eventually establishing controls and reducing the role of nuclear weapons in world politics.

If the superpowers have the right to hold potential proliferators to account for the likely effects of their actions on third parties, so also do third countries have the right to hold the superpowers accountable in similar terms. The non-proliferation regime imposes obligations on the superpowers, which they have not adequately fulfilled. Even though the regime does not rest primarily on Article 6 of the NPT, if the regime collapses, the superpowers will be to blame for the consequences to the extent that their failure to meet obligations undercut it. Given the importance of alliance guarantees in stemming proliferation, however, such obligations under Article 6 cannot be interpreted as simple disarmament. And such alliances have a moral dimension in helping smaller nations to preserve their independence without adding to the general risks that arise from proliferation.

What would be the consequences of the spread of nuclear weapons to more countries? One cannot say for certain. In part the answer depends on "who" and "when." It is possible that stable deterrence could be created in some regions. But, for the reasons stated above, the prospects are uncertain at best. Given the fears that new nuclear capabilities among neighbors create, there will be a temptation to destroy them by preemptive strikes while they are still small and vulnerable. The country that expects to create stable deterrence in a region by introducing nuclear weapons may have to pass through a dangerous "valley of vulnerability." Even when stable deterrence is imaginable in a region, it may be highly risky to try to get from here to there. Moreover, the rate of spread makes a difference. A slow rate makes it more possible to manage the destabilizing effects and reduce the risk of nuclear use than it would be in a situation where nuclear weapons are spreading quickly. But given the dangers of increased risks described above, it is wise to err on the safe side. In any event, there is a good moral case to be made for a policy of non-proliferation, but it is important to remember that the obligations bind in two directions.

# CHAPTER SEVEN

---

# Moral Choices and the Future

GIVEN THE ENORMITY of the potential effects, moral reasoning about nuclear weapons must pay primary attention to consequences. In the nuclear era a philosophy of pure integrity that would "let the world perish" is not compelling. But given the unavoidable uncertainties in the estimation of risks, consequentialist arguments will not support precise or absolute moral judgments.

Those two propositions have several important implications for moral reasoning about nuclear weapons. First, sound ethical conclusions will be contingent on empirical facts and hypothetical arguments about strategic interactions as well as values. They will not be derived solely from simple principles. This requires that those wishing to make good moral arguments become informed about a complex and often uncertain set of empirical facts. Effective moralists will have to dirty their hands with some knowledge of nuclear strategy. Second, the range of nuclear issues that can be settled by appeal to moral principles alone will be restricted. Most specific issues relating to nuclear weapons will turn upon empirical, strategic, and prudential arguments rather than solely on moral principles. The first step in moral reasoning about nuclear weapons should be not to expect too much. Once the moralists enter the realm of contingency, they must tread carefully.

A third implication is equally important. Given the inevitable uncertainties in simple consequentialist reasoning, arguments should not rest solely on that dimension. The uncertainties surrounding nuclear calculations are too great to put complete faith in simple consequentialism. Calculation of consequences should be supplemented by rules or maxims derived from historically tested doctrines like just war theory. On nuclear issues, as in other issues discussed in Chapter 2, good moral reasoning is three-dimensional and includes considerations of motives and means. Considering consequences is a necessary but not a sufficient basis for sound moral reasoning about nuclear weapons.

Some moralists would go to the other extreme and reject the necessity of the consequentialist approach, which they find too dependent on complex and difficult arguments about what the world is actually like. "People who are concerned about nuclear weapons often prefer to take what one might call the high road of morality. They feel terror and revulsion about the whole thing; they sense that the strategic arguments will lead them further into the bog."[154] Others fear that strategic arguments will suppress "the indignation that ought to form part of any human being's reaction to the nuclear age."[155] Consequentialist thinking, in this view, is part of the problem, not part of the solution. But such a view is stunted moral reasoning.

## *Nuclear Abolitionism*

Antinuclear moralists often draw analogies to such historical precedents as the antislavery movement and contend that it was necessary to use absolutist arguments and arouse public indignation in order to abolish that evil institution. Similarly, the public will have to change its way of thinking and be roused to indignation if we are to abolish nuclear deterrence. But indignation can be costly in the nuclear age. If it is not aroused equally in separate sovereign states, it may merely increase anxiety without solving the basic problem. Alternatively, a misshapen, asymmetrical crusade can lead to disastrous consequences.

The slavery analogy raises troubling considerations about nuclear consequences when one remembers that abolition took more than a century to accomplish, split a nation, and helped to precipitate a costly war. And equally to the point, slavery is a misleading analogy for nuclear deterrence. Unlike Ramsey's babies on bumpers, slaves were being subjected to a hurtful experience very different from the citizen's experience of risk under a policy of nuclear deterrence. Moreover, at least in the democracies, the majority of citizens have expressed consent to their hostage status. In short, the slavery analogy is both factually confusing and morally misleading.

Policies derived from such abolitionist analogies might increase rather than diminish nuclear dangers. For example, after he contemplated "double death" in *The Fate of the Earth*, Jonathan Schell concluded that it would be necessary to "reinvent politics." He decided that humankind would have to abolish the national sovereignty of separate states. When critics argued that neither was likely in the near term and that therefore he had no solution that fitted with his apocalytic indignation, Schell proposed disarmament and "weaponless deterrence" as a deliberate policy in *The Abolition.* But such a "policy" is so poorly supported by Schell's reasoning about its consequences that it is not a real solution to the moral dilemma. In fact, it could raise risks of nuclear war rather than reduce them. Since Schell is among the most provocatively interesting of the abolitionist writers, a careful look at his argument is a useful way to illuminate some of the difficulties with nuclear abolitionism.

Schell proposes "weaponless deterrence," where "factory deters factory; blueprint deters blueprint," and so on. We can obtain the benefits of deterrence at zero levels of weapons instead of at the level of 50,000. Each country will know that if it rearms, others will do likewise and all will be worse off. If one country does nonetheless rearm, it will be weeks or months before it can strike, and that will allow other countries to prepare to respond. There will be more time and less destruction if deterrence fails under such a system than under the present system of nuclear deterrence.

How can this be accomplished? Schell gives a four-point plan: (1) Existing nuclear powers will agree to abolish their arsenals, and nonnuclear countries will agree not to acquire such weapons. (2) Countries will agree to limit and balance the conventional forces that would now be of critical importance in the absence of nuclear weapons. (3) After the abolition of offensive nuclear weapons, antinuclear defense systems would be added to stabilize the situation. (4) Each country would be allowed to keep its ability to rearm with nuclear weapons at some agreed level, such as a six- or eight-week lag. That would be the final guarantee of weaponless deterrence.

What is wrong with such a plan? Nothing, except how

94

little relationship it has to the world we live in. If policy formulation means adjusting ends and means in a world of constraints, this is not an exercise in "deliberate policy." For example, on Schell's first point, how are we supposed to get the abolition agreement? What has changed that has made it less elusive? Schell argues that enforcement rests on the knowledge that breakdown would be to no one's advantage. But what if some states decide to cheat? For example, why has a state like Libya signed the Non-Proliferation Treaty yet covertly continued to seek nuclear weapons? Would such behavior cease just because other countries signed an abolition agreement? And given the problems discussed in the previous chapter, would proliferation of near nuclear capability really be stabilizing in such a world, as Schell argues it would be?

Regarding Schell's second point, at least he recognizes that the abolition of nuclear weapons will increase the importance of conventional weapons. But how likely are countries to agree on equal limits? And given the geographical proximity of the Soviet Union to key areas of Europe and Japan, would equal limits be adequate to ensure their protection against their giant neighbor? If not, how would that affect the global balance of power and our foreign policy interests? How would the Soviets react? Not only are we not told, but Schell's book is virtually innocent of any discussion of the Soviet Union.

As for the antinuclear defenses that Schell advocates in his third point, how likely is it that countries will wait until after the abolition or severe reduction of offensive forces before developing such defenses? If they do not, why won't one side's concern about the other side's improving defenses lead it to increase its offensive weapons to preserve its deterrent capability? How does Schell propose to avoid such an arms race? And will not the defense have to be perfect if abolition is to be tolerable?

Finally, Schell asserts that a world in which all states were six or eight weeks away from reinventing or reconstructing nuclear weapons would "mark a revolution in stability." But would it? What if a secretive society like the Soviet Union were able to shorten its lag to two weeks? Or suppose we just believed that to be the case—as we believed in the spurious missile gap in the past? Would there not be the worst kind of arms race to fabricate crude nuclear devices without the safety features such as electronic combination locks that currently protect them? And would there not be a tremendous temptation to fit nuclear warheads on the dual-capable conventional delivery systems as quickly as possible and to strike preemptively at the other side's nuclear factories and labs laden with blueprints? Rather than a deterrent, those factories would become vulnerable lightning rods for a first strike!

Schell scoffs that "that idea, for example, the Soviet Union using a clandestine nuclear force, could destroy the ability of the United States to make nuclear weapons and then, in the space of a few weeks, conquer Europe, cross the Atlantic, and occupy the United States to prevent nuclear rearmament is patent fantasy." But why cross the Atlantic? After the first two steps, the Soviet Union would be in possession of Europe and could threaten continued intercontinental preemption or worse if we refused to accept the *fait accompli* and attempted to rebuild our nuclear facilities that it had destroyed.

The inadequacy of abolitionist analogies and arguments does not mean that one must accept what nuclear strategists offer as the final word in consequentialist arguments. For one thing, strategists disagree with each other, and behind apparently dispassionate analysis there is often a mixture of self-interest and unrealistic assumptions. Many strategists treat deterrence like a religion or ideology in which dogma is invoked to support preferred conclusions, and dominant schools of thought anathematize heretical thinkers.[156] Detailed elaborations of unrealistic scenarios are used to support expensive programs with little sense of proportion or attention to competing moral claims. But the best way to respond to inadequate strategic thinking is not simply with moralistic indignation but with better consequentialist analysis.

## Maxims of Nuclear Ethics

The standards for good moral reasoning about nuclear weapons are those described in Chapter 2, including clarity, logic, consistency, and attention to unnoticed consequences, as well as an awareness of all three dimensions of ethical judgment: motives, means, and consequences. Those standards can help provide general maxims for moral reasoning about nuclear weapons. They do not resolve all specific issues, but they allow us to go further than invoking Michael Walzer's doctrine of "supreme necessity" or the European Catholic bishops' "logic of distress," which would have the effect of asking us to drop our moral principles at the door before we enter the domain of nuclear thinking,[157] when what we really need are some moral maxims that allow us to find a path through the confusing thickets of the nuclear forest.

If we deny ourselves the self-indulgence of absolute judgments regarding nuclear weapons, then like the European and American Catholic bishops, we shall have to regard the moral acceptability of nuclear deterrence as conditional. For the French bishops, the four critical conditions are: (1) deterrence applies only to self-defense; (2) the avoidance of overarmament; (3) precautions against mistakes; and (4) a constructive policy that serves the cause of peace. For the American bishops, the three

97

key conditions are: (1) continually to say no to nuclear war; (2) rejection of the quest for superiority; and (3) deterrence should be used as a step on the way toward progressive disarmament.[158] The American bishops then go on to recommend a number of specific proposals (such as no-first-use, a test ban, deep cuts, and removal of short-range nuclear weapons). For that they have been accused of wedding theology to a particular theory of deterrence. Some of their specific proposals are debatable in terms of their likely effects both on deterrence and on the prospects for avoiding nuclear war.[159] But one need not agree with all the detailed judgments (which are more contingent on facts) in the different national statements to see the outlines of useful principles in their overall approach.

Like the just war tradition that they draw upon, the bishops pay heed to all three dimensions—ends, means, and consequences—in reaching their overall judgment. The French bishops are particularly good on the strength and limits of just cause. An important contribution of the American bishops is their introduction of a temporal dimension when considering consequences. One of the greatest problems in consequentialist arguments is a failure to distinguish measures appropriate for the short and long terms. What is good for the long term may be disastrous in the short term, and vice versa. For example, premature efforts to abolish nuclear weapons in the short run might raise risks of nuclear use, but efforts to reduce reliance on nuclear weapons over the long term may make better sense. By considering deterrence as a step toward some indefinite longer term, the American bishops introduce a useful distinction.

In sum, following the bishops' examples and drawing upon the broad just war tradition, we can conclude that nuclear deterrence is conditionally moral, the three conditions being: (1) a just cause that is proportionate to the likely means and consequences; (2) limits on means; and

(3) a prudent consideration of consequences in both the near term and the indefinite long term. When we look at what those three conditions imply for a moral approach to our nuclear dilemma, we find, as indicated earlier, that those moral conditions do not resolve all specific issues. On the other hand, they do suggest a few maxims or simple commandments to keep in the back of one's mind to help bring moral reasoning to bear in an appropriate way when one is evaluating particular proposals or situations. The first three maxims relate to motives and means in a revised just war tradition as I discussed in Chapter 4. The last two maxims relate to short- and long-term consequences and obligations among generations, which were discussed in Chapter 5.

## *Five Maxims of Nuclear Ethics*

MOTIVES

1. Self-defense is a just but limited cause.

MEANS

2. Never treat nuclear weapons as normal weapons.

3. Minimize harm to innocent people.

CONSEQUENCES

4. Reduce risks of nuclear war in the near term.

5. Reduce reliance on nuclear weapons over time.

By becoming more specific, or by separating out the issue of competing moral claims, one could generate a longer list of principles. But the greater the specificity, the greater the contingency. And the intuitions we must rely upon at times of action are more strongly formed if the principles are fewer and more general rather than multiple and more contingent. There can be certain tensions between the maxims, for example, between 2 and 3 or between 4 and 5, but as we shall see below they also reinforce each other. The whole is greater than the sum of the parts.

## 1. Self-Defense Is a Just but Limited Cause

Except for pacifists, self-defense is widely accepted as a legitimate cause for doing harm to those who threaten us with imminent harm. And survival is not the only value we wish to defend. If we treat it as such, we create a self-fulfulling prophecy. We may have life and nothing else, and perhaps not even that. Red or dead are not the only alternatives. (Given the fact that China and the Soviet Union have come close to war in the past two decades, it may be possible to be both red and dead!) But the moral argument rests on the fact that self-defense includes more values than merely biological survival. We are also justified in protecting our culture and freedoms, which give value to our lives beyond mere existence.

As I argued in Chapter 5, the acceptance of risk to preserve values other than survival does not do an injustice to future generations. On the principle of equal access to values, and given the fact that there is no harm to identifiable individuals (since they do not yet exist), there is no reason to assume that future generations would make survival an absolute value and not wish us to take proportionate risks to ensure their access to cultural and political freedoms.

Self-defense, however, is an elastic and ambiguous concept in international politics, particularly in a world with a high degree of nuclear bipolarity underlying its balance of power. Some conclude that self-defense must be interpreted literally to exclude anything other than deterrence of a nuclear attack (for some, any attack) upon the territory of the United States. They would reject the legitimacy of the goal of extending our nuclear deterrence to protect other countries from nuclear (or for some, conventional) attack. Those who reject extended deterrence argue that their position is more clearly consistent with the moral principle of self-defense. Moreover, they might add that since a Soviet attack against the American homeland is one of the less likely paths to nuclear war, the rejection of extended deterrence could have the consequence of lowering risks of nuclear war.

But that rejection of extended deterrence is almost certainly a spurious conclusion. The danger of harm to the homeland may increase later because of an earlier failure to maintain a balance of power. And while the balance of power is a somewhat ambiguous concept, and equilibrium involves more than merely power politics, the equilibrium since 1945 has rested on an American alliance with two of the five great centers of industrial power—Europe and Japan—even though they are geographically closer to the Eurasian centers, Russia and China. Failure to maintain those alliances would at the very least drastically change that equilibrium and lead to unintended consequences, some of which might increase rather than diminish the risks of nuclear war. Among the negative consequences of the withdrawal of extended deterrence would be a probable increase in the proliferation of nuclear weapons among some of the countries now sheltered by the American nuclear umbrella. Moreover, it is of no small moral consequence that the Western alliance is an alliance of democracies based on a process of consent and with the effect of protecting critical values of political freedom. The argument against extended deterrence fails if one allows pooled self-defense. Whether or not that could ever be restricted to conventional weapons depends on empirical and strategic judgments. But in terms of values, the argument against extended deterrence fails to establish strong consequential reasons to believe that the goal of preserving the political freedoms of Western democracies is disproportionate to the likely results of trying to pursue it. Indeed, one might argue that running some risks to protect a billion people in democracies is more moral than merely protecting 230 million.

At the same time one should be aware of the limits of arguments based on extended deterrence. Extended deterrence is not universal global deterrence. The credibility of extended deterrence depends more heavily on the relative stakes that the two countries have in a region than on the military balance. Those stakes may depend more on trade and culture than on weapons.[160] And conventional weapons and the local conventional balance may be more important than nuclear weapons in the complex mixture of relative stakes, conventional weapons, and nuclear weapons that determines the credibility of extended deterrence. In other words, extended deterrence cannot legitimately be used as an excuse for simply piling up more and more nuclear weapons when there may be alternative and safer means to achieve the same goal. And extended deterrence must be seen as a matter of degree that tapers off as stakes become lower. That means no amount of nuclear weapons will make extended deterrence credible in areas where we have low stakes (e.g. Afghanistan, Angola), and nuclear threats in such circumstances would be foolish and risky bluffs.

Finally, it is worth noting that different countries will take different views of the appropriate mix of nuclear and conventional weapons in extended deterrence, depending on their different situations. Europeans realize that if deterrence fails, the ensuing battle—whether conventional or with tactical nuclear weapons—will destroy the place where they live. They tend to prefer deterrence that includes threats of early escalation to Soviet (and American) homelands. Americans tend to look for firebreaks or principles that will forestall or delay the use of nuclear weapons.

Some Americans argue that nuclear weapons have no role in deterring conventional war; they can be used only to deter nuclear attack. As an empirical proposition, that lacks *prima facie* plausibility. As a longer term goal it may make sense as a way of reducing the role of nuclear weapons, but as a moral declaration it glides too quickly over the different situations of our allies and the difficult question of sharing the burden of risk. Sometimes ambiguous formulas like "flexible response" and "no early use," which are the result of bargaining among democratic allies, have better consequences than ringing (but less credible) declarations like "no first use." While there is long-term merit in moving toward a position in which we do not plan for first use of nuclear weapons, the overall consequences of a declaration that nuclear weapons should only deter nuclear attack would be good in the current circumstances only if it were freely accepted among all the allies.[161]

Are there other causes besides individual and pooled self-defense that justify nuclear deterrence? Perhaps, but there are serious problems to be overcome in admitting them. Good motives are not enough. Other causes must also pass the moral tests of impartiality and proportionality. In the thought experiment described in the previous chapter, it was possible to imagine other (nonallied) nations agreeing that some would be allowed a right of nuclear self-defense as long as their goals were limited and the resulting nuclear balance provided a means of preserving order as an alternative to chaos and high risk of loss of sovereignty. But it is unlikely that other motives would be admitted in a world of divergent cultural interpretations of basic values among states with a desire to preserve their independent cultures.

Too often Americans justify actions to maintain the nuclear balance of power with the argument that our motives are good. But such arguments ignore the problems of hubris and cultural self-centeredness, which may lead others to see our well-intended acts as overbearing or imperialistic. Moral argument includes a respect for the perspectives of others. Even though it is appropriate to argue that Western moral values are better, the French bishops correctly warn us against the dangers of cultivating a "Manichean conception of the world: all the evil on one side, all the good on the other."[162] A crusade to rid the world of totalitarianism or Communism or to make the world safe for democracy (or human rights as we understand them) might seem admirably motivated to us but hopelessly self-centered to others, and far too risky to accept in a world of nuclear weapons. While other nations may accept some nuclear risks imposed upon them for the sake of self-defense and minimal order, they are unwilling to see those risks seriously raised by a global crusade for American-defined values. Not only might they perceive the values and their implementation differently, but they would almost certainly regard the risks as too high to allow the justification of the proportionality of the goal.

In sum, just cause is limited to self-defense (including pooled self-defense), and a foreign policy that rests upon nuclear deterrence must include prudent restrictions on its policy objectives.

2. Never Treat Nuclear Weapons as Normal Weapons

One can construct and aim a nuclear weapon that is so small and accurate that it will do about the same damage as conventional "iron bombs" full of high explosives. But even those miniature nuclear weapons must never be treated as normal usable weapons, because politically and technically they are too closely related to their big brothers of mass destruction.

At the same time, as I argued in Chapter 4, we have to consider the usability dilemma. If nuclear weapons are regarded as completely unusable, they lose the deterrent effect that is their *raison d'être*. And some solutions, such as a demonstration shot in the Artic, may suggest weakness that tempts an aggressor rather than a resolve that deters. Thus credible targeting seems necessary for deterrence. Yet the effort to identify credible military targets has raised a heated debate over the legitimacy of planning for "nuclear war-fighting" as opposed to nuclear deterrence. In one sense this debate is spurious, for some planning for delivery of nuclear weapons against military targets is both planning for "war-fighting" and a necessary means of deterrence. But to the extent that planning and discussion of prevailing in war after repeated massive nuclear strikes encourages "notions that nuclear war can be engaged in with tolerable human and moral consequences," it is malicious in its effects.[163] Declaratory policy affects perceptions and procedures, and, given the potential immoral consequences, it is important to maintain perceptions and procedures that do not treat nuclear weapons as normal usable weapons.

It is possible that a nuclear war can be kept limited, but the risks of failure are too great to count on a high likelihood. Careful planning for precise limited strikes is an appropriate exercise in limiting military means if a political leader is to have any choices between capitulation and an all-out spasm of total devastation. And the existence of such choices is an important component of the credibility of deterrence. But political (and military) leaders must beware the danger of false precision and never forget the abnormality of actually making such a choice. A century ago Bismarck remarked that when you draw the sword, you roll the dice. When the dice are nuclear, prudence shades into a moral maxim.

Moreover, leaders must be wary of creating machinery that operates as if nuclear weapons were normal regardless of original intentions. Deterrence is not a game between individuals; it is played among small groups in large, complex organizations. If weapons are tightly integrated with forces and military exercises are continually conducted as though they will be used, not only will nonnuclear defense atrophy, but the presumption in favor of nuclear use may grow on both sides. Remembering our nonrational model of the onset of nuclear war discussed in Chapter 5, it is important that the view of nuclear weapons as abnormal be built into bureaucratic standard operating procedures (daily storage, exercises, release procedures) as well as kept in mind by political and military leaders.

The danger that things will get out of control has been used by some people to suggest that credible uses may not be as important to deterrence as the rational actor model suggests. The very existence of nuclear weapons creates deterrence. Given their potentially horrendous consequences, one does not need a very high probability of use to create *inherent* or "existential deterrence."[164] The deterrent effect arises not from calculation, but from uncertainty; from the prospect that deliberately or not, nuclear weapons might be used. It is inherent in their very existence. That is an important point often neglected by those who promote elaborate schemes for usable weapons as the only way to enhance deterrence. In a sense, one can conceive of nuclear deterrence as consisting of concentric circles based on fear of the different ways by which a nuclear war might start. The hard core of deterrence rests on the nonrational model and the fact that war by inadvertence will always be a looming specter encouraging prudence in a world where the dice are nuclear. The outer circle of deterrence is credible options for use of nuclear weapons that encourage prudence in the calculations of a rational opponent.

The relative significance of the two dimensions of deterrence—inherent and calculated—is often debated, and the answer depends in part on one's view of the relative importance of the two models of how a war might start. But even those who are more concerned about inadvertent war and who place more reliance on inherent deterrence need not believe that it would be prudent to rely on inherent deterrence alone. If we did, we would be celebrating the role of accident and reducing the role of human agency—hardly a happy moral position. And inherent deterrence must be related in some way to the calculated strategic actions we are trying to deter. If we put all our weapons in salt mines or relied on quickly reassembling them under "weaponless deterrence," they would have little deterrent effect. To compensate for the very low probability of use by increasing the horrendous consequences, for instance by building some sort of doomsday machine, would be neither prudent nor moral.

Deterrence is bound to continue to be based on a mixture of calculated and inherent considerations. So long as that is true, the usability dilemma will remain. And so long as that dilemma remains, it will be important to protect against unlimited means by convincing leaders, the public, and other nations that nuclear weapons must never be treated as normal weapons. Plans for escalation control may exist, but they should be regarded as plans for limiting damage if deterrence fails. They should not be considered a realistic basis for a strategy of threatening gradual application of nuclear force in battle. The prospects for control are too uncertain to allow us to use such threats to substitute for the more costly task of improving conventional forces. There is no justification for taking disproportionate risks, particularly when alternative means are available.

## 3. Minimize Harm to Innocent People

In principle we wish to avoid harm to all people, but on the grounds that allowed us to reject the pacifist objection to all killing, we have a special obligation to the innocent. Since innocent people pose no threat, we cannot justify killing them under the moral principle of self-defense. But since nuclear deterrence requires some prospect of use, and use may get out of control, deterrence will always impose some risk of harm on innocent people. As Russell Hardin has pointed out, even unilateral disarmament would impose a risk of harm on innocent people in this country.[165] Steps that seriously weaken deterrence merely shift risks among innocent people and may raise risks to all. Ever since nuclear weapons were invented there has been some unavoidable prospect of harm to innocents, but a number of things can be done to reduce rather than increase the risks borne by the innocent. In the revised nonabsolute form suggested by O'Brien and discussed in Chapter 4, minimizing harm to innocents is a way to meet at least some of the concerns expressed in the just war tradition.

We can reduce the deliberate killing of noncombatants by avoiding or limiting the targeting of cities. Not only is that relevant to children in Soviet society, but since the burning of cities is a necessary condition for potential nuclear winter, it would reduce the harm to innocents in third countries as well. If deterrence should fail, it is important that it be a small failure and that leaders have time to regain their senses rather than a preset program for massive slaughter of innocents.

Current strategic plans do not deliberately include cities, but given the location of many military, industrial, and communication targets within or near cities, the distinction fades into insignificance. A serious effort to avoid cities in strategic planning would require us to forgo a number of communication and industrial targets, which are given a high priority in the countervailing and warfighting doctrines currently fashionable. Ironically, some of the harshest critics of the immorality of the doctrine of Mutual Assured Destruction (MAD) would produce quite similar effects in their harm to innocent people.[166]

Advocates of Mutual Assured Destruction argue that the exclusion of cities would not be desirable. Since some threat to innocents is unavoidable in any case, and the whole point of nuclear targeting is to avoid a breakdown of deterrence, the very horror of counter-city attacks reinforces deterrence. The knowledge that a nuclear attack would destroy virtually the entire urban population produces great prudence. It destroys the temptation of precise counterforce attacks with their illusions of escalation control. More important, it provides a specific upper limit on the number of nuclear weapons needed for deterrence, and that limit can be quite low. Thus it is preferable in terms of competing moral claims on resources. If 400 weapons (two submarines' worth) can destroy half of Soviet society, then the rest is overkill. We could cut our strategic arsenal by 90 percent and still have enough to deter the Soviet Union. And since innocent people are harmed not only by a chance of being killed in nuclear war but also by the diversion of scarce resources from competing moral claims such as health and education, there is a good case to be made for finite deterrence.[167]

In contrast, counter-silo targeting can lead to an infinite spiral of increasing weapons or considerable fears of instability in a crisis. If there are single-warhead missiles, each side must add two missiles to be sure of destroying each new silo the other side builds (or enough to barrage large areas where mobile missiles would roam.) On the other hand, if there are multiple-warhead missiles in fixed silos, there is an enormous premium for striking first in a time of deep crisis. When opposing weapons are very vulnerable and very lethal, they give rise to the fear that one must "use them or lose them." An attack against some silos may lead the attacked side to discharge weapons ("launch under attack") that otherwise might be withheld.

Despite those merits, the overall case for large-scale assured destruction targeting is not convincing. Its potentially suicidal nature reduces its credibility for calculated deterrence. And in the case of inadvertent war, it bears much the same faults as a doomsday machine (which, according to the nuclear winter scenarios, it is.) If war breaks out by accident, the last thing we want is to be programmed for holocaust. Moreover, the benefits of restricting the strategic arsenals to low levels may be partly illusory if it is true that 100 megatons on each side targeted against cities might bring on nuclear winter. Finally, as a matter of integrity, there is something morally unacceptable about planning a massive slaughter of Soviet children when there are other ways to accomplish deterrence. A critical consequential question is how well deterrence can be accomplished without the massive targeting of cities. If that is impossible, the effect of a no-cities doctrine would merely be to shift or raise risks among different innocent people.

There is a long history of dispute between advocates of counterforce and of countervalue targeting.[168] Some of the dispute is spurious. MAD became a declaratory measure of the adequacy of forces in the mid-1960s primarily as a means of creating finite limits in the strategic budget. It never became the basis of the targeting plan. And those who attacked MAD in the 1970s in order to urge the development of new, prompt counter-silo weapons failed to distinguish between the *doctrine* of assured destruction targeting and the *condition* of ultimate vulnerability that remains even when the doctrine is changed. Many of their scenarios paid too little attention to the inherent or existential dimension of deterrence.

In fact, one can imagine alternative targeting doctrines that would avoid the problems of large-scale assured destruction targeting while also avoiding the problems of counter-silo targeting.[169] One way to ease the dilemma would be by a limited counter-city doctrine; another would be a counter-combatant rather than a counter-silo approach to counterforce targeting. Each approach would escape the constrictions of the standard debate summarized in Figure 2A.

Generally speaking, counterforce discussions have focused on silos, and countervalue discussions have assumed

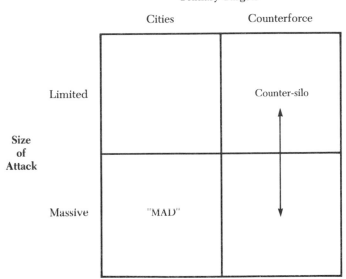

FIGURE 2A.
*Current Nuclear Targeting Debates*
**Primary Targets**

FIGURE 2B.
*Alternative Nuclear Targeting Doctrines*

**Primary Targets**

| | Cities | Armies | Silos |
|---|---|---|---|
| Limited | "Tit-for-tat" countervalue | Counter-combatant counterforce | |
| Massive | | | |

Size of Attack

massive destruction of cities. Neither is necessary. As Figure 2B shows, one can imagine limited counter-city targeting as well as counterforce strategies that do not focus on silos.

The problem with targeting silos is prudential. The appearance or reality of a first strike may lead the other side to launch weapons they might otherwise withhold. A counter-military or counter-combatant targeting doctrine would avoid the excesses of prompt hard-target counterforce versus counter-city choices.[170] Targeting nuclear weapons to destroy advancing Soviet armies (and the bridges, railheads, and narrow points though which they must pass and the supply depots they must rely upon) in the West and South, and Soviet border defense in the East, could deny the Soviet Union the gains from any aggression while exposing it to additional dangers. Innocent people would be killed in such attacks, but far fewer than in a massive counter-city strategy. Many of those counter-military targets are already included in the current strategic plans; others would require new investments in intelligence and targeting capabilities. But if

more attention were given to minimizing harm to innocents, those options may become more attractive and
more feasible.

Counter-combatant targeting doctrines have the virtue
of reducing harm to civilians and reducing the risks of
nuclear winter. But they require forces larger than the
small forces proposed by advocates of finite deterrence
(so beloved of philosophers who write about nuclear
weapons), for the requirements of destroying Soviet armies and border defenses are large and could increase
over time. On the other hand, such a doctrine would
establish somewhat clearer and lower limits than the current emphasis on counter-silo and counter-industrial targeting does.[171] If combined with the prudent approach
to nuclear use suggested in Maxim 2 (rather than worst-
case planning to cover all conventional targets with strategic nuclear weapons), it need not lead to a massive
arms race.

There are two other objections to a counter-combatant
targeting doctrine. Some fear that such a doctrine will
make weapons appear too usable and lead us to treat
them as normal. But the uncertainties and dangers of
escalation will place limits on that confidence. Again, the
second maxim—never treat nuclear weapons as normal—
must be kept in mind. Others fear that leaving cities as
sanctuaries will seriously weaken deterrence. They might
argue that a robust counter-combatant deterrent must
include the capability to destroy at least part of a city
where the Soviet leadership might be cloistered in a time
of crisis. Indeed, if deterrence failed and we suffered a
massive strike against our cities, it would make more
sense in the last resort to be able to retaliate only against
Soviet leadership, not all Soviet children.[172] Such a limited
exception may be justified, but in addition the uncertainties provide a hard core of existential deterrence. Soviet
leaders know that weapons can be quickly reprogrammed

to cover many targets, and cities are the easiest targets to strike. There is little reason to believe that a no-cities doctrine (perhaps with a limited exception regarding top leadership) would seriously weaken deterrence.

Thomas Schelling has suggested a limited counter-city targeting approach as an alternative way to ensure deterrence and reduce harm to innocents.[173] A limited "tit for tat" attack on as few cities as possible could produce the prudence and limits on the number of weapons that are the benefits of MAD, while avoiding the pitfall of being massively preprogrammed for holocaust. Such an approach would be like a disaggregated doomsday machine. If such a doctrine provided a more effective deterrent at a level that allowed a substantial reduction in weapons, it might reduce harm to innocents and reduce nuclear weapons in comparison with the current approach.

In consequential terms, both those alternative targeting doctrines are better than current approaches, but it is difficult to determine which of the two alternatives will do more good over the long term. A key question is whether graduated counter-city attacks would be harder to keep under control than graduated counter-combatant attacks would be. What we know about Soviet doctrine suggests that would be the case, but we cannot be certain. A second question is, if control were lost under either strategy, whether there would be a difference in the harm done to noncombatants before the war were brought to an end.[174] Perhaps less harm would be done by the counter-combatant approach. Both those questions about consequences are difficult to answer. In situations where consequential analysis is so uncertain, we might make the choice by invoking our sense of integrity about reducing the direct threat of harm to innocents, preferring a counter-combatant targeting doctrine. Ultimately, deterrence in both cases rests in part on the hard core of the possibility of destroying populations, but doing harm to innocents is not as imminent or essential to a counter-combatant strategy. The injunction to minimize harm to innocents—whether it be by direct targeting of civilians, by the threat that nuclear winter poses to third countries, or by the diversion of scarce resources from competing moral claims—should encourage us to explore new approaches to strategic doctrine that will be more morally defensible in democratic societies. But the ethical problems of nuclear deterrence cannot be solved by adjustments of targeting doctrine alone.[175] No targeting doctrine is perfect, and minimizing harm to innocents depends even more on reducing risks of nuclear war—both deliberate and inadvertent—in the first place.

4.  Reduce Risks of Nuclear War in the Near Term

Since it is hard to find anyone in favor of nuclear war, the fourth maxim seems obvious until one remembers that some risk of nuclear war is necessary for deterrence and that even unilateral disarmament involves some risk of nuclear wars. How to lower risks and avoid nuclear war is not as obvious as it might first appear. It depends in part on how nuclear war might start.

As I argued above, there are three general views of the onset of nuclear war. Hawks emphasize calculated deterrence based on rationality and argue that the best way to avoid nuclear war is to strengthen our deterrent posture, that is, to maintain the survivability of our retaliatory forces. Even the risk of some overinsurance in the numbers of weapons is a small cost compared to the failure of deterrence. Doves point out that in circumstances of extreme threat constrained rationality may suggest preventive or preemptive attack, and reassurance of

115

one's adversary by reducing armaments would be a more appropriate way to reduce risks.

Both positions have abstract merit, but where one finds the appropriate balance between deterrence and reassurance depends, in part, on Soviet perceptions of opportunity, risk, and risk aversion (and Soviet perceptions of our perceptions). Strategic interaction is a two-party game, and knowing our own assessment of relative risks is only half the story that determines outcomes. The appropriate balance between deterrence and reassurance would be very different if one imagined a nuclear-armed Hitler. That is only probable, not certain, and it may depend on the opportunities we are perceived as presenting to our opponents. The good news is that there is little evidence to support the view that the current Soviet leadership has a Hitlerian approach to war and risk, and popular analogies between the current period and the 1930s are historically flawed. The bad news is that much remains unknown to us about Soviet perceptions.[176] For reasons deeply rooted in Russian history and Communist ideology, the Soviet Union remains a secretive and repressive society, which affords us little direct access to the perceptions of their elites.

That problem of perceptions has bedeviled Western discussions of nuclear strategy and efforts to decide how many weapons are enough for deterrence.[177] On one view, symmetry is irrelevant. Analogies and wisdom from the prenuclear era are misleading. All we need is a fixed number of invulnerable weapons able to do some unacceptable damage to the Soviet Union in any conceivable circumstances. We can then relax and let them waste their money on overkill. In the alternative view, symmetry is crucial, because it is Russians, not Americans, who we are trying to deter.[178] We can infer what they consider to be militarily and politically important by the size and nature of the forces they deploy. We need a similar structure to deter better, and thus reduce the risks of nuclear war. In addition, if we allow the Soviet Union to gain a significant advantage in numbers of weapons (or some significant category such as land-based missiles), they may misread our failure to respond as signifying a shift in the vague Marxist notion of the correlation of forces. That may embolden them to take risks that could raise the probability of nuclear war as a consequence of our reaction to an unacceptable threat to some vital national interest.[179]

It is virtually impossible to settle this difference definitively, given the inadequacy of our current knowledge of the Soviet military and political leadership and decision processes. Nonetheless, one can set certain limits. Since symmetry is much more expensive in terms of competing moral claims, and since arguments about perceptions are so intangible and subject to abuse, a somewhat larger portion of the burden of proof should be shouldered by those arguing for symmetry. Such arguments will often prevail, but given both sides' acceptance of certain asymmetries in the past, one should be skeptical of justifications of expensive new systems or arguments based solely on perceptions. Such systems should also be judged in terms of their effects on the risk of nuclear war by nonrational causes or by a mixture of the rational and nonrational models. For example, it does not follow from the fact that the Soviets have developed a first-strike capability against our land-based missiles (a quarter of our strategic warheads) that we should develop a similar capability against theirs (which constitute nearly three-quarters of their strategic warheads). One needs to weigh the gains for calculated deterrence (and avoidance of war) from such symmetry against the increased risk of instability at a time of deep crisis. Owls point out that such tradeoffs can change dramatically in a time of crisis. Since the fear of a disarming first strike can have a devastating effect on stability in a time of crisis, avoiding the appearance of such a strategic capability is an important criterion for reducing risks of nuclear war.

Indeed, the Owl's approach to the onset of nuclear war warns us against trying to gain small increments in calculated deterrence based on rationality (the Hawk's assumption) when there are substantial costs in terms of the risks of nonrational onset or inability to terminate nuclear war (the Owl's assumption). A healthy respect for uncertainty and skepticism about overly complex rational strategies would lead one to reject certain policies as violating the moral maxim to reduce risks. For example, a policy of launching our missiles on warning would help to deter a Soviet first strike (hawkish assumptions) but at the cost of greatly increased vulnerability to catastrophic failure (on owlish assumptions). A strategy for decapitating the Soviet leadership in Moscow might enhance deterrence (Hawk's view) but might produce chaos that might make war termination impossible if deterrence ever failed (Owl's view). Delegating authority over battlefield nuclear weapons to field commanders might enhance deterrence (Hawks) but might also make war termination impossible (Owls). Using nuclear alerts for political signaling might help establish credibility in a crisis (Hawks), but it would also relax the safety procedures that guard against accidental nuclear war (Owls) in exactly the situation where safety catches may be most needed. Hawkish horizontal escalation theories in which we respond at the point of our choosing ("Cuba for Iran") may sound like a rational strategy in normal times, but increasing the number of simultaneous crises might create unmanageable owlish problems in coping with psychological stress and bureaucratic control under nuclear crisis conditions. One of the recurrent temptations for nuclear strategists is to be too clever by half. Elaborate strategies are developed solely on the basis of calculations

FIGURE 3.
*Balanced Deterrence Avoids Three Errors*

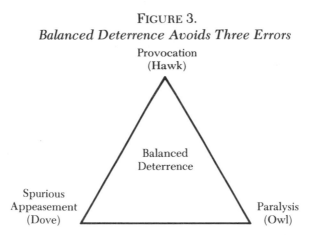

Provocation
(Hawk)

Balanced
Deterrence

Spurious
Appeasement
(Dove)

Paralysis
(Owl)

based on rationality. But much of human history has turned upon unanticipated disaster.[180] In such a world, owlishness is a necessary condition for consequentialist moral reasoning.

At the same time, owls do not have a monopoly on wisdom. In terms of policy for avoiding nuclear war, Hawks, Doves and Owls each have part of the truth, but not the whole. Each camp's prescriptions taken to a logical extreme could lead to fatal error of provocation, temptation, or paralysis, as Figure 3 illustrates.

In deterrence situations, one must beware the Hawk's hair trigger, the Dove's dropped gun, and the Owl's frozen safety catch. What one needs to avoid nuclear war and reduce risks is a policy of balanced deterrence that minimizes the prospects of all three errors. What that means in practice will involve detailed empirical and strategic argument. At the same time, such arguments are related to moral reasoning. A key problem in consequentialist moral reasoning about nuclear weapons is assessing the risks posed to critical values and establishing a degree of proportionality among them. Facts and values are closely intertwined. Some broad principles are suggested in Table 3 and are spelled out in detail elsewhere.[181] The

TABLE 3
*Principles for Avoiding Nuclear War
in the Near Term*

1. Maintain a credible nuclear deterrent
2. Improve conventional deterrence
3. Enhance crisis stability
4. Reduce the impact of accidents
5. Develop procedures for war termination
6. Prevent and manage crises
7. Invigorate non-proliferation efforts
8. Limit misperceptions
9. Pursue arms control negotiations

first two are hawkish and the last is dovish, but the large majority are owlish.

Measures to reduce risk are seldom costless, and such costs in terms of resource diversion, foreign policy, and political process must be carefully weighed. But it is important to see nuclear risk reduction and the avoidance of nuclear war in the near term as a moral imperative flowing both from considerations of proportionality between means and ends, and from the imperative to buy time for the development of better alternatives over the long term.

## 5. Reduce Reliance on Nuclear Weapons over Time

Since humans are fallible, it seems reasonable to assume that nuclear deterrence may someday fail. But as I argued in Chapter 5, that is not a certainty, and metaphors based on constant probabilities (such as flipping a coin) are misleading. If we can continually reduce probabilities, failure is not inevitable in any meaningful time frame. Moreover, even if the failure of deterrence were "inevitable," this proposition does not necessarily entail the apocalyptic conclusions that are often drawn from it. Failures need not be large ones that lead to nuclear winter and extinction of the species. Indeed, though it is not to be wished, a small failure may turn out to be a way to innoculate against a larger one. And even if a large failure is inevitable, that inevitability may come after thousands of years. What we do now can affect those outcomes. By pursuing prudent objectives and reducing risks (Maxims 1 and 4), we can reduce the likelihood of failure. By treating nuclear weapons as abnormal and minimizing harm to innocents (Maxims 2 and 3), we can reduce the probable size of the failure if it eventually occurs. From the proposition that deterrence may fail someday it simply does not follow that nuclear holocaust is either imminent or inevitable. As Richard Garwin has pointed out, "it is not really true that a nuclear war is inevitable. If the probability of nuclear war this year is one percent, and if each year we manage to reduce the probability to only 80 percent of what it was the previous year, then the cumulative probability of nuclear war for all time will be 5 percent."[182]

Nonetheless, any chance of holocaust or species extinction is so horrible that we would be wise to reduce our reliance on nuclear weapons as much as possible in a way that does not raise the current risks of nuclear deterrence's failing.[183] Given the fallibility of human reason and human organizations, and the inevitable spread of nuclear technologies to more countries (and possibly terrorist groups), it would seem wise gradually to reduce our reliance on nuclear weapons. Whether that can ever lead to abolition of weapons is impossible to say. Certainly it would be impossible to abolish nuclear knowledge without burning all books and all scientists. The prospect for that solution may have passed when the Pope failed to burn Galileo! Even if the knowledge of nuclear weapons is an inescapable part of the trajectory of human history, there is nothing inevitable about the number of weapons and the centrality of their role in military and political relations among nations. Seen in those terms, Richard Ullman is correct to suggest the wisdom of a gradual denuclearization of world politics (even though one need not accept all his specific proposals to reach that goal).[184]

Robert Tucker argues that "the first and, indeed, the last thing that needs to be said in any moral assessment of deterrence, is that it should not be regarded as a temporary arrangement. . . . Its justification now cannot be made to turn on a prospect that from the vantage point of the present must appear as near utopian. . . . Deterrence, it appears, will remain a part of our 'condition' for as far as we can presently see."[185] Tucker is correct to caution us against overly optimistic attempts to escape deterrence in the near term, for such utopian thinking could have highly immoral unintended consequences. But he would be wrong if he meant to discourage long-term utopian thinking. We simply know too little about the distant future to make such categorical assertions, and a little dose of utopian thinking now about the distant future may involve few short-term risks while helping us to set our policies in a direction where solutions that are currently improbable become possible at some time in the future.

Almost by definition, the current probabilities of utopian solutions are low, but they may become higher for future generations. We simply do not know. But we should avoid foreclosing such prospects and should take commensurate risks to keep such possibilities open. Trying gradually to reduce our reliance on nuclear weapons in world politics is such a proportionate risk.

Some people describe the problem of going beyond nuclear deterrence as one of defining a perfect solution; a disarmed world and a world government are common examples. It is not difficult to imagine such worlds, but it is much harder to think realistically of how to create them. And it is not always clear that they would be better.

If disarming all nuclear weapons removed a deterrent of chemical and biological warfare or if a world government capable of controlling high-technology violence meant a repressive and intolerant empire, we might regret our utopia. But rather than imagining some ideal (or horrid) end point in an unknowable future, we might encourage "process utopianism" and try to imagine ways to consistently lower the risks of nuclear war even if we are not sure of the end point. The idea of building peace or embedding nuclear deterrence in an improving context may be more fruitful than the usual "end point utopianism." And the development of a nuclear risk reduction strategy is not an impractical idea.

How might one envisage moving toward less reliance on nuclear weapons and lower risks of nuclear war in the future? Two centuries ago Immanuel Kant suggested that peace among nations might come about as a result of three things: the increasing destructiveness of war; the spread of republican governments; and the growth of commerce and trade. In the short run of two centuries, Kant seems to have been proved wrong. Extensive trade and enormous destructive capability did not prevent Europe from destroying itself in World War I. But perhaps a century and a half was too short to test the validity of Kants' prescriptions. Unlike 1914, when many leaders (and their citizens) glorified and desired war, nuclear devastation has made war far less attractive today. And while the time and number of cases may not be large enough for us to treat the evidence as conclusive, it is interesting that there are no significant instances of war in this century among republican (that is, democratically elected representative) governments.[186]

An effort to predict long-term change is likely to be frustrated, but one can sketch rough outlines of broad paths of change that might lead to a future with less reliance on nuclear weapons. Roughly speaking, there are technological paths and social-political paths. The latter are the more promising, but the former are currently receiving more attention.

Among the most dramatic technological changes of the past decade has been the increase in the accuracy with which weapons can be delivered. That has reduced the nuclear destructive power needed to destroy particular targets and has increased the precision with which some military targets can be separated from civilian ones. In addition, there have been important improvements in the identification of changing targets and the flexibility of the weapons that might attack them. Those changes cut two ways. On the one hand they permit greater discrimination and the possibility of counter-combatant targeting; on the other hand they give rise to the temptation to think of nuclear weapons as militarily useful in battle. For reasons given above, that would be a serious mistake. But the same improvements in accuracy and identification of targets can be used to improve the capabilities of conventional weapons and allow them to replace battlefield nuclear weapons. In other words, if wisely applied, the technology of accuracy could help to reduce reliance on battlefield nuclear weapons and on large, inaccurate strategic weapons that would destroy entire cities. But technology alone will not solve our ethical dilemmas. To argue that "the *live* issue is whether we should be trying to increase or to decrease our ability to discriminate between military and civilian targets and to confine destruction to the military" is insufficient.[187] It may be correct on the rationalistic assumptions that underlie calculated deterrence, but is not an adequate response to the nonrational factors that "Owls" believe would lead to the failure of deterrence. Technological improvements that improve discrimination in targeting help to alleviate some moral dilemmas, but as I argued earlier, targeting doctrines will not alone provide adequate solutions to issues of nuclear ethics. They must be combined with measures to ensure control and reduce risks of nuclear war. Such an approach reaches beyond technology and deep into psychology and politics.

The technological promise that has received greatest attention in recent years is the prospect of effective defense against ballistic missiles. President Reagan has called it a "moral imperative." But the language of morality has been bandied about loosely and inappropriately by both sides in the so-called star wars debate. It is a good example of stunted moral reasoning in the discussion of specific nuclear issues. President Reagan has expressed a desire to escape from the dilemmas of nuclear deterrence; he suggested in his speech of March 1983 that strategic defense might provide such an exit. But to escape from deterrence requires a leakproof defense not only against ballistic missiles but also against bombers and cruise missiles. Such a perfect defense seems unlikely. Some who doubt the technological feasibility of such a task have urged that the new Strategic Defense Initiative Organization in the Pentagon concentrate on the lesser task of defending missiles rather than cities.[188] The rationale would be to enhance rather than replace deterrence.

But that is no more a moral imperative than are alternative ways of enhancing deterrence. While a case can be made (and debated) for such defense, it has to surmount the obstacles of feasibility and cost in terms of competing strategies and competing moral claims on resources. It is not moral simply because of good intentions. Feasibility is more than a matter of technology.[189] It also has a political dimension. Even if the complex defense technology can be developed and combined into a system that has to be perfect without ever having been tested under the stress of nuclear war, can we get from here to there without going through a transition period in which the nuclear predicament of both sides would be worse? Will improving defenses stimulate a massive increase in offensive weapons, as the other side tries to prove it can still overcome the defenses? And if it looks as though we are about to perfect a defensive system in a few years that would effectively disarm the Soviet deterrent, would Moscow take more risks during crises in the short run because it saw the likelihood that those risks will pay off politically diminishing with the passage of time? Can those problems really be solved by offering to share the technology, when one realizes that the key technologies of computers and sensors also provide the base for our improvement of the conventional force balance in Europe?

Those questions do not mean that there should not be a significant research program on strategic defense. Quite the contrary. The prospect of enhancing deterrence, and perhaps in the long term of being able to save a large number of lives, justifies some research effort. But the morality of such an initiative will depend on the consequences, not the motives. It may be "better to defend than to avenge," but only if the consequences of *trying* to defend do not increase the risk of nuclear conflict in the meantime. And those consequences are likely to be determined not by our intentions but by the type of technology chosen (how it will affect crisis stability) and by the political state of U.S.–Soviet relations (Will the introduction of defense occur in a cooperative or an antagonistic setting?).

The political and social paths lack the glamour of the technological one, and, as many scientists have sadly discovered, politics is harder than physics. But political and social change is the key factor in the long-range future, and political change can take a number of forms. There may be changes in relations between the United States and the Soviet Union; there may be changes in the growth of international institutions and cooperation among states; there may be changes in domestic political and social attitudes toward the sovereign state and its defense.

Sometimes political relations between states can change quite quickly—witness the change in relations between the United States and China in less than a decade. But such rapid changes are often related to the existence of a common enemy, and it would be unwise to expect such rapid change in U.S.–Soviet relations.

There are several deep-seated reasons to expect tension in U.S.–Soviet relations. First, as Tocqueville already saw in the nineteenth century, the enormous size and resources of the Russian and American nations foreshadowed a future bipolar rivalry. Then, in 1917 the Bolshevik Revolution created a layer of deep ideological incompatibility. When World War II destroyed the multination balance of power existing before 1939, it left a bipolar structure of world power centered on the U.S.–Soviet rivalry. The accumulation of vast nuclear arsenals overshadowing those of all other nations has consolidated that "special relationship." From a power politics point of view, the probability of tension is built into the very structure of the relationship.

Nonetheless, at different times there have been lesser degrees of hostility. And it is worth remembering that the worst outcome has not occurred. Despite hostility, there has also been prudence in managing the world's first nuclear balance of power. The destructiveness of nuclear weapons introduced a disproportion between most ends that the superpowers seek and the principle military means at their disposal.[190]

That situation has led to the evolution by a process of trial and error of some primitive rules for avoiding or managing crises. The rules are so primitive that they might more correctly be called prudent practices. Indeed, they began well before the onset of détente in the 1970s and survived its demise. As described by Stanley Hoffmann, "one such informal rule was the non-resort to atomic weapons. . . . A second rule was the avoidance of direct military clashes between their armed forces. . . . A third element was the slow (and for America) painful learning of limited wars . . . calculated so as to limit

127

the risks of escalation, even if those constraints made a clearcut victory or a rapid settlement impossible. Later came the beginnings of nuclear arms control between Washington and Moscow."[191] Rudimentary as those rules or prudent practices are, they are significant if one believes that a stable balance of power requires a degree of moderation in the actors' behavior, as well as a military balance.

Where there is a degree of common interest in stability, the balance of power can become a positive situation in which both sides win. Indeed, an international "regime"—a set of tacit or explicit rules and procedures—may be developed to encourage a stabilizing perspective of long-range rather than short-range self-interest. The nineteenth-century balance of power and the contrast between Bismarck's restraint and the failure of his successors are often cited as an example.

To say that the postwar U.S.–Soviet balance of power has been embedded in such a regime would be to stretch a point. The rules are ambiguous and not openly accepted by both sides. In 1972 and 1973 Nixon and Brezhnev signed agreements that seemed to codify the rules, but the ambiguities (such as exceptions for "wars of national liberation") later led to a sense of cheating and deception after the Middle East War of 1973 and the Soviet transport of Cuban troops to Angola in 1975 and 1976. Moreover, an agreed regime implies reciprocity and flexibility in bargaining behavior as the opponents seek to avoid jeopardizing the regime. In U.S.–Soviet relations, however, reciprocity has proved to be limited in time and on issues. It is difficult to "bank goodwill" from one time or issue to another in the relationship. While both sides have a degree of interest in preserving stability in the form of preserving nuclear bipolarity and the avoidance of nuclear war, their competition makes it impossible to agree on the *status quo*. While a common interest exists, it is severely limited and strained by the competitive dimension.

Theorists of international cooperation point out, however, that warm relations and trust are not necessary for cooperation.[192] Competition does not prevent cooperation, and a central authority is not necessary. What is needed are a more sophisticated and long-range perception of self-interest, an ability to learn from experience, and the realization that a relationship will continue over time. Various experiments show that when games of competition and cooperation are played over long periods, a strategy of reciprocity proves to be most effective. When the shadow of the future looms large over the present, the incentives for cooperation between competitors are strengthened.[193]

On such realistic premises, it is quite possible to expect a gradual evolution of U.S.–Soviet cooperation. Certainly relations in the 1980s are vastly different from those in the 1950s. Despite frequent allegations of return to the Cold War, there are far more contact and cooperation in the 1980s than there were thirty years ago.[194] There is far greater awareness of each others' perception of its vital interests, and of its internal processes. It is highly plausible to expect change in the relationship over the next thirty years and more.

In addition, there may be changes inside the Soviet Union. The Soviet Union is unlikely to change quickly, but it does evolve slowly and unevenly. The Soviet Union has opened up somewhat over the last thirty years. There are more contacts. There is more sophistication in the perception of outside reality. There are more pinholes letting light into the "black box."

Change inside the Soviet Union will be gradual and hard to gauge, and we can at best encourage it rather than hope to guide it. Nonetheless, the possibility suggests that a managed balance of power and nuclear risk reduction strategy should involve routine and regular communication. Our approach should include engaging the Soviets in prolonged strategic discussions; holding talks at a high level on force structure and stabilization measures; and efforts to consider crisis prevention techniques, not necessarily in the expectation of signing formal agreements, but as a means of enhancing transparency and communication. It also follows that trade, scientific and cultural exchanges, and tourism should be evaluated by the same standard and not solely by the current criteria of economic benefits or short-term security interests. A managed balance of power strategy does not rest on expectations that increasing engagement will win Soviet trust or greatly constrain Soviet actions. Nor does it rest on the hope of any immediate liberalizing effects. In the first instance, it rests on the importance of enhancing mutual transparency and communication.

Beyond the U.S.–Soviet relationship there may be other political changes that will affect the role of nuclear weapons in world politics. That the development of cooperation among states rests on self-interest rather than goodwill is encouraging, because it implies that cooperation is consistent with realistic premises. There may continue to be a gradual development of international regimes that govern various dimensions of economic and social interdependence among states. In some cases those regimes may involve a gradual growth of institutions. In other instances, economic integration among states may reduce the degree of potential conflict as it has between ancient enemies like France and Germany. And the development of transnational institutions and contacts may gradually transform domestic attitudes toward sovereignty and the use of violence to defend the state. The development of transnational interests, organizations, and communication may encourage the growth of multiple loyalties that soften the claims of the sovereign state and of groups that help to bridge the cultural parochialisms of different peoples.[195] After all, the modern territorial state, which replaced looser feudal loyalties, has been the dominant institution only since the Peace of Westphalia in 1648. It is not implausible to think that it will give way to other institutional forms if we look forward for an equivalent period. But one need not imagine replacing the nation-state; it is enough to draw its fangs. In a world of diverse peoples separate political communities help to preserve valuable differences. Instead of imagining the world's peoples getting into one big boat of world government, it may be better to think in terms of lashing the many national boats more firmly together.[196]

Of course, this Kantian view of the distant future is only one among a vast number of possible outcomes. Other benign (or horrendous) outcomes are also possible. The important point is that we should not let our imaginations be captured solely by images of imminent nuclear holocaust or by cynical views of the immutability of our dependence upon nuclear weapons, for both images tend to stifle the modest efforts we can and should take now to make a Kantian type of future somewhat more probable for future generations to develop. As discussed in Chapter 5, our moral obligations to future generations are to avoid large risks now—of either war or the sacrifice of freedoms—and to try to ensure their choices by trying gradually to reduce our reliance on nuclear weapons whenever we can do so without unacceptably increasing current risks. And our obligation to our own generation is to be explicit about our values and careful in our moral reasoning about those critical choices.

### Conclusions

No one knows how likely a Kantian type of future is, and the farther out we try to look, the less we can see. But that is all right, for future generations will have their own values and must make their own choices. We owe future generations what we owe ourselves: a chance to enjoy both freedom and survival by pursuing such objectives with a proportionate (that is, modest) degree of risk. I have suggested five moral maxims for how to approach our nuclear dilemma. This approach overlaps with prudential and strategic argument, but it differs as well. I have drawn those maxims from both the personal-integrity-oriented and the consequentialist traditions of Western moral philosophy, and have addressed all three dimensions—motives, means, and consequences—that are essential to good moral reasoning. The approach avoids the common strategists' trap of pretending amorality while implicitly smuggling unexamined values into their choices. It also avoids the moral absolutists' trap of valuing their personal integrity so highly that they are unconcerned with consequences even when they include the possibility "that the earth may perish." And it avoids the abolitionists' trap of promoting solutions that might make sense in the long run but may cause great pain and unnecessary disaster in the near term.

The maxims of nuclear ethics suggested here do not attempt to solve all nuclear dilemmas. On the contrary, there is no way to avoid complex empirical arguments and often enormous uncertainty. But these maxims of "just deterence" can give leaders a basis for informing their intuitions when they have to make decisions in situations where there is little time to think through one's moral philosophy. And the maxims may give citizens in our democracy a basis for judging and reacting to the rough outlines of suggested policies as well as a sense of hope and moral effectiveness rather than a pessimism that fosters pain or complacency. There are ways to bring our Western traditions of moral reasoning to bear on the central dilemma of our time, and there are good reasons to believe that "the first generation since Genesis" can live in freedom without being the last.

# Notes

1. United States Catholic Conference, "The Challenge of Peace: God's Promise and Our Response," *Origins* (National Catholic Documentary Service), 13 (May 19, 1983): 1.
2. Jonathan Schell, *The Abolition* (New York: Knopf, 1984), pp. 5, 77.
3. Michael Walzer, *Just and Unjust War* (New York: Basic Books, 1977), p. 282.
4. *The New Republic,* December 20, 1982, p. 7.
5. The Public Agenda Foundation, *Voter Options on Nuclear Arms Policy* (New York: The Public Agenda Foundation, 1984).
6. Charles Krauthammer, "The End of the World," *The New Republic,* March 28, 1983, pp. 12–15.

7. Fred Kaplan, *The Wizards of Armageddon* (New York: Simon & Schuster, 1983), pp. 232–33.

8. Colin S. Gray, "Strategic Defense, Deterrence, and the Prospects for Peace," *Ethics*, 95 (April 1985): 661.

9. See Robert Kennedy, *Thirteen Days* (New York: Norton, 1968), and John Lewis Gaddis, "The Postwar International System: Elements of Stability and Instability," unpublished paper.

10. *New York Times*, May 6, 1983, p. A32.

11. Marshall Cohen, "Moral Skepticism and International Relations," *Philosophy and Public Affairs*, Vol. 13, Fall 1984; see also Charles Beitz, *Political Theory and International Relations* (Princeton, N.J.: Princeton University Press, 1979).

12. Arnold Wolfers, *Discord and Collaboration* (Baltimore: Johns Hopkins Press, 1962), Chapter 4, "Statemanship and Moral Choice."

13. Stuart Hampshire, ed., *Public and Private Morality* (Cambridge: Cambridge University Press, 1978), Chapter 2.

14. *Time*, July 8, 1985, p. 21. Or in the words of William Colby and Stansfield Turner, other former directors of the CIA, "assassination is beyond what our country can stand," and "we are a country that embodies a lot of legal principles as well as moral ones."

15. Freeman Dyson, *Weapons and Hope* (New York: Harper & Row, 1984).

16. The quote is from the British philosopher Michael Dummett, "Nuclear Warfare," in Nigel Blake and Kay Pole, eds., *Objections to Nuclear Defence* (London: Routledge & Kegan Paul, 1984), p. 33.

17. For example, René Louis Beres argues that "it is difficult to rule out schizophrenia as a possible explanation of the Reagan Administration's strategic mythmaking." "Vain Hopes and a Fool's Fancy," *Philosophy and Social Criticism*, 10, No. 3/4 (1984): 44.

18. Robert J. Lifton and Richard Falk, *Indefensible Weapons* (New York: Basic Books, 1982), p. ix.

19. For a sophisticated presentation of this view, see Susan Khin Zaw, "Morality and Survival in the Nuclear Age," in Blake and Pole, *Objections to Nuclear Defence,* pp. 115–43. Zaw justifies the emotivist approach on the grounds of "a false separation between morality and survival" and an overly bleak view of a "senseless future" that requires us to "change the world" (p. 134). I explain why this view is too bleak in Chapters 5 and 7.

20. Dorothy Austin, "Reinventing Ourselves: The Spiritual Task of the Nuclear Age," *Political Psychology,* 6, No. 2 (1985): 333, 326.

21. See Alasdair MacIntyre, *After Virtue* (South Bend, Ind.: University of Notre Dame Press, 1981).

22. S. I. Benn, "Persons and Values: Reasons in Conflict and Moral Disagreement," *Ethics,* 95 (October 1984): 37.

23. Jonathan Glover, *Causing Death and Saving Lives* (Harmondsworth, Middlesex: Penguin, 1977), p. 25.

24. Quoted in *New York Times,* July 9, 1982, p. 22.

25. Derrick de Kerckhove, co-director of the McLuhan Program, University of Toronto, quoted in *New York Times,* September 30, 1984, p. 41.

26. Sidney Hook, Letter to the Editor, *New York Times,* June 19, 1984, p. 26. Hook went too far, however, by calling the possibility of nuclear death "merely speculative." Strictly speaking, it is an unknown probability.

27. R. M. Hare, *Moral Thinking* (Oxford: Clarendon Press, 1981), Chapter 10; see also J. L. Mackie, *Ethics* (Harmondsworth, Middlesex: Penguin, 1977), p. 192.

28. Amy Gutman and Dennis Thompson, eds., *Ethics and Politics* (Chicago: Nelson-Hall, 1984), p. xii.

29. These terms are similar to Max Weber's distinction between an "ethics of conviction" and "an ethics of responsibility". W. D. Hudson, *A Century of Moral Philosophy* (London: Lutterworth, 1980) distinguishes the two traditions as intuitionism and consequentialism. Robert Gordis points out another distinction between an ethics of absolute love and self-abnegation and an ethics of justice or

self-fullfillment, both implicit in the New Testament, by noting that the golden rule does not require us to love our neighbors *more* than ourselves. "Religion and International Responsibility," in Kenneth Thompson, ed., *Moral Dimensions of American Foreign Policy* (New Brunswick, N.J.: Transaction, 1984).

30. There are, of course, important differences within each tradition. Consequentialism includes both "act utilitarians" and "rule utilitarians," as well as some who believe that the concept of utility is too narrow and imprecise to measure all consequences. The person-centered ethics of virtue includes both the deontological or rule-oriented Kantian approach and the broader aretaic or Aristotelian approach. And that does not exhaust the distinctions! But for our purposes here, the difference between the two broad approaches is sufficient. See Bernard Williams, *Morality* (New York: Harper & Row); William Frankena, *Ethics* (Englewood Cliffs, N.J.: Prentice-Hall, 1973); and Tom L. Beauchamp, *Philosophical Ethics* (New York: McGraw Hill, 1982).

31. There are important differences in agent-centered theories between Aristotelians and Kantians, with the latter judging integrity more narrowly in terms of the will to act in conformity with rules derived from pure reason. For Aristotelians, integrity means embodying a number of virtues or traits. For Kantians, an action "has moral worth only when it is performed by an agent who possesses what Kant calls a 'good will'; and a person has a good will only if moral duty based on a universally valid norm is the sole motive for the action." Beauchamp, *Philosophical Ethics* (Note 30), p. 117. See also Frankena, *Ethics* (Note 30).

32. One might distinguish the two traditions as an "ethics of good" versus an "ethics of right." The problem with such a distinction, however, is that common usage has blurred the two terms. While a purist may reserve the term "morality" for perfect duties within a system of fundamental or categorical restraints, many people use the

word "moral" (and the associated words "right," "ought," and "just") in the broader sense described in this chapter. One may regret that the purity of the words has been lost, but it seems odd to argue, as Terry Nardin does, that "contemporary usage notwithstanding, it seems to me mistaken to refer to utilitarianism or other consequentialist theories as moralities." "Nuclear War and the Argument from Extremity," in Avner Cohen and Steven Lee, eds., *Nuclear Weapons and the Future of Humanity* (Totowa, N.J.: Littlefield, Adams, 1984).

33. Paul Ramsey, *The Just War: Force and Political Responsibility* (New York: Scribner's, 1968).

34. The following paragraph is an adaptation of a hypothetical case constructed by Bernard Williams in J. J. C. Smart and Bernard Williams, *Utilitarianism: For and Against* (Cambridge: Cambridge University Press, 1973), p. 98.

35. For an interesting discussion, see Sissela Bok, "Kant on the Maxim 'Do What is Right Though the World Should Perish,' " in David Rosenthal and Fadlou Shihadi, eds., *Theoretical and Applied Ethics* (forthcoming). Nardin, "Nuclear War and the Argument from Extremity," argues that the maxim "expresses the confidence of the ages that in fact the world will *not* come to an end" and that "the world is endangered not by mankind's fidelity to these traditions but by its rejection of them." Perhaps that is true, but from a consequentialist perspective it is not very comforting.

36. Norman Podhoretz, *Why We Were in Vietnam* (New York: Simon & Schuster, 1982).

37. See David Hollenbach, S.J., "Ethics in Distress," paper presented at Georgetown University Conference on Justice and War in the Nuclear Age, March 1983.

38. Walzer, *Just and Unjust War* (note 3), Chapters 16 and 17.

39. William H. Shaw, "Nuclear Deterrence and Deontology," *Ethics,* 94 (January 1984): 253.

40. Steven Lee, "The Morality of Nuclear Deterrence: Hos-

tage-Holding and Consequences," *Ethics*, Vol. 95, April 1985. Consequentialism is sometimes defined as the ethical doctrine that only consequences matter. Strictly speaking, that remains true of a broad consequentialism that incorporates rules about means or motives, but the substantive differences in the approaches begin to diminish.

41. Douglas P. Lackey, *Moral Principles and Nuclear Weapons* (Totowa, N.J.: Rowman & Allanheld, 1984), p. 5.
42. William Safire, *New York Times*, November 19, 1984, p. 23.
43. For example, "start and end with rules": Treat rules as *prima facie* obligations and calculate whether the consequences justify overriding that obligation, but if there is great uncertainty in the calculations and/or the expected values turn out to be roughly even, decide on the basis of rules.
44. Robert Osgood and Robert Tucker, *Force, Order, and Justice* (Baltimore: Johns Hopkins Press, 1967), p. 252.
45. Hampshire, *Public and Private Morality* (note 13), p. 53.
46. "Introduction," *Ethics*, Vol. 95, April 1985.
47. See Charles Beitz, "Bounded Morality," *International Organization*, 33 (Summer 1979): 405–24. My categorization differs by distinguishing between skeptics and realists.
48. Hans J. Morgenthau, *Politics Among Nations* (New York: Knopf, 1958), p. 9.
49. John Rawls, *A Theory of Justice* (Cambridge: Harvard University Press, 1971), p. 378.
50. Michael Walzer, "The Moral Standing of States," *Philosophy and Public Affairs*, 9 (Spring 1980): 211.
51. Stanley French and Andres Gutman, "The Principle of Self-Determiniation," in Virginia Held, Sidney Morgenbesser, and Thomas Nagel, eds., *Philosophy, Morality, and International Affairs* (New York: Oxford University Press, 1974).
52. Gerald Doppelt, "Statism Without Foundations," *Philosophy and Public Affairs*, 9 (Summer 1980): 401–3.
53. David Luban, "The Romance of the Nation State," *Philos-*

*ophy and Public Affairs,* 9 (Summer 1980): 392; see also Robert Amdur, "Rawls's Theory of Justice: Domestic and International Perspectives," *World Politics,* April 1977, pp. 438–61.

54. Stanley Hoffmann, *Duties Beyond Borders,* (Syracuse: Syracuse University Press, 1981), p. 155.

55. Walzer correctly points out (in private correspondence) that the sophisticated application of his state moralist position can lead to similar judgments in particular cases. Since any sophisticated approach must come to terms with the significance of both states and individuals, the differences of approach may often be more in the starting points than in where they come to rest between states and cosmopolitanism in particular cases.

56. For a challenge to this distinction, see Peter Singer, *Practical Ethics* (Cambridge: Cambridge University Press, 1979).

57. Michael Walzer, "The Distribution of Membership," in Peter Brown and Henry Shue, eds., *Boundaries* (Totowa, N.J.: Rowman & Allanheld, 1981), p. 1.

58. See for example, Charles Beitz, "Cosmopolitan Ideals and National Sentiment," and Henry Shue, "The Burdens of Justice," *The Journal of Philosophy,* Vol. 80, No. 10 (October 1983).

59. See the further critique of Rawls in Michael J. Sandel, *Liberalism and the Limits of Justice* (Cambridge: Cambridge University Press, 1982).

60. See Alasdair MacIntyre, "The Magic in the Pronoun 'My,'" *Ethics,* 94 (October 1983): 123.

61. Gerard Elfstrom, "On Dilemmas of Intervention," *Ethics,* 93, No. 4 (July 1983): p. 711.

62. For examples see Robert O. Keohane and J. S. Nye, "Transgovernmental Relation and International Organization," *World Politics,* Vol. 57, October 1974.

63. See Richard Cooper, "A New International Economic Order for Mutual Gain," *Foreign Policy,* 26 (Spring 1977): 81 ff.

64. For elaboration, see J. S. Nye, *Ethics and Foreign Policy* (Wye, Md.: Aspen Institute, 1985).

65. Ends are closely related to but not always synonymous with motives. Some actions may be motivated by goalless or expressive or irrational behavior—by pushes as well as pulls. I am indebted to Henry Shue for discussion of this point.

66. See Glover, *Causing Death and Saving Lives* (note 23), pp. 255–58; Lackey, *Moral Principles and Nuclear Weapons* (note 41), pp. 9–11; and David Hollenbach, *Nuclear Ethics: A Christian Moral Argument* (New York: Paulist Press, 1983), Chapter 2.

67. Quoted in Hollenbach, *Nuclear Ethics,* p. 66.

68. See U.S. Catholic Conference, "Challenge of Peace" (note 1), p. 9.

69. Pope John Paul II, quoted in *New York Times.* May 18, 1985, p. 3.

70. Comment at the European–American Conference on Ethical Aspects of Nuclear Deterrence, Ebenhausen, April 19–21, 1985.

71. James Turner Johnson, *Can Modern War Be Just?* (New Haven: Yale University Press, 1984), p. 175.

72. William V. O'Brien, "The Challenge of War: A Christian Realist Perspective," in Judith A. Dwyer, S.J., ed., *The Catholic Bishops and Nuclear War* (Washington, D.C.: Georgetown University Press, 1984), pp. 42, 44. See also Michael Novak, *Moral Clarity in the Nuclear Age* (Nashville: Thomas Nelson, 1983).

73. Barrie Paskins, "Deep Cuts Are Morally Imperative," in Geoffrey Goodwin, ed., *Ethics and Nuclear Deterrence* (New York: St. Martins, 1982), p. 115.

74. Charles Krauthammer, "On Nuclear Morality," in James Woolsey, ed., *Nuclear Arms: Ethics, Strategy, Politics* (San Francisco: ICS Press, 1984).

75. U.S. Catholic Conference, "Challenge of Peace," p. 23; Catholic Bishops of France, "Winning the Peace," in James V. Schall, ed., *Out of Justice, Peace: Winning the*

*Peace* (San Francisco: Ignatius Press, 1984). See also Francis X. Winters, S.J., "After Tension, Detente: A Continuing Chronicle of European Episcopal Views on Nuclear Deterrence," *Theological Studies*, Vol. 45, 1984.

76. Johnson, *Can Modern War Be Just?*, p. 177, and Thomas Donaldson, "Nuclear Deterrence and Self-Defense," *Ethics*, April 1985.

77. Ralph Potter, *War and Moral Discourse* (Richmond: John Knox Press, 1969), p. 51.

78. Joseph C. Kunkel, "Right Intention, Deterrence, and Nuclear Alternatives," *Philosophy and Social Criticism*, 10, No. 3/4 (1984): 145. Or as Paul Ramsey put it, although there may be a right of reprisal in international law, the fact that an enemy destroys our society first "cannot make it any less of an injustice for us to destroy his." "The Limits of Nuclear War," in Thompson, *Moral Dimensions of American Foreign Policy* (note 29), p. 140.

79. Harvard Nuclear Study Group, *Living With Nuclear Weapons* (Cambridge: Harvard University Press, 1983), p. 248.

80. Arthur Hockaday, "In Defense of Deterrence," in Goodwin, *Ethics and Nuclear Deterrence*, p. 71.

81. Barton Gellman, "The Weinberger–Thompson Debate," *The American Oxonian*, 81 (Spring 1984): 115–19.

82. President Reagan, quoted in *Time*, February 18, 1985, p. 19.

83. Robert W. Tucker, "Morality and Deterrence," *Ethics*, 95 (April 1985): 477.

84. Germain Grisez, "The Moral Implications of a Nuclear Deterrent," *Center Journal* (Winter, 1982), p. 16. See also Schell, *The Abolition* (note 2), p. 77.

85. Schell, *The Abolition*, p. 56.

86. See the assumptions, for example, in Richard Wasserstrom, "War, Nuclear War, and Nuclear Deterrence: Some Conceptual and Moral Issues," *Ethics*, April 1985, which argues that "given the structure and design of policies of nuclear deterrence, no mistakes or accidents are tolera-

ble because one such failure—no matter what the cause—eradicates the system and all of those individuals ostensibly served by it.

87. Francis X. Winters, "The Nuclear Arms Race: Machine vs. Man," in Harold P. Ford and F. X. Winters, eds., *Ethics and Nuclear Strategy* (Maryknoll, N.Y.: Orbis, 1977), p. 151.

88. Donaldson, "Nuclear Deterrence and Self-Defense" (note 76), p. 541.

89. Aspen Strategy Group, *The Command and Control of Nuclear Weapons* (Wye, Md.: Aspen Institute, 1985). See also Desmond Ball, "Can Nuclear War Be Controlled?" Adelphi Paper No. 169 (London, IISS, 1981); John Steinbruner, "Launch Under Attack," *Scientific American*, 250, 1 (January 1984); and Bruce G. Blair, *Strategic Command and Control* (Washington, The Brookings Institution, 1985).

90. See Edward Luttwak, "How to Think About Nuclear War," *Commentary*, 74, 2 (August 1982).

91. See Steven Meyer, "Soviet Perspectives on Paths to Nuclear War," in G. Allison, A. Carnesale, and J. Nye, eds., *Hawks, Doves, and Owls* (New York: Norton, 1985).

92. See J. Bryan Hehir, "The Context of the Moral-Strategic Debate and the Contribution of the U.S. Catholic Bishops," paper delivered at Princeton University, April 1984, p. 18., and Albert Wohlstetter, reply to critics, *Commentary*, December 1983, p. 17.

93. Catholic Bishops of France, "Winning the Peace" (note 75), p. 109.

94. Robert S. MacNamara, "The Military Role of Nuclear Weapons," *Foreign Affairs*, Vol. 62, Fall 1983.

95. William Shaw, "Deterrence and Deontology," *Ethics*, January 1984, p. 252.

96. Hardin, "Deterrence and Moral Theory" (note 31).

97. Gregory Kavka, "Some Paradoxes of Deterrence," *The Journal of Philosophy*, 75 (June 1978): 291.

98. Bernard Williams, "How to Think Sceptically About the Bomb," *New Society*, November 18, 1982, p. 289.

99. Germain Grisez, "Chapter 2: Moral Doctrine No Longer at Risk," *National Catholic Register,* April 24, 1983.

100. See, for example, Jonathan Bennett, "Morality and Consequences," in James P. Sterba, ed., *The Ethics of War and Nuclear Deterrence* (Belmont, Calif.: Wadsworth, 1985). Moreover, some Catholic moralists worry about the meaning of "direct" in the doctrine. See Richard McCormick, S.J., "Notes on Moral Theology," *Theological Studies* 46 (1984): 51 ff.

101. Kaplan, *Wizards of Armageddon* (note 7), pp. 211–12.

102. William V. O'Brien, "Just War Doctrine in a Nuclear Context," *Theological Studies,* 44 (June 1983): 211.

103. Donaldson, "Nuclear Deterrence and Self-Defense" (note 76).

104. Hardin, "Deterrence and Moral Theory"; Tucker, "Morality and Deterrence" (note 83).

105. Kavka, "Some Paradoxes of Deterrence," p. 301.

106. Russell Hardin, "Risking Armageddon," unpublished paper.

107. See Schell, *The Abolition* (note 2), and Lackey, *Moral Principles and Nuclear Weapons,* (note 41).

108. Lifton and Falk, *Indefensible Weapons* (note 18).

109. Paul Schroeder, "Does Murphy's Law Apply to History? *The Wilson Quarterly,* Winter 1985, p. 88.

110. See Gaddis, "The Postwar International System" (note 9).

111. Thomas Powers, *Thinking About the Next War* (New York: Knopf, 1982). p. 17.

112. Todd Gitlin, "Time to Move Beyond Deterrence," *The Nation,* December 22, 1984, p. 676.

113. Schell, *The Abolition,* p. 22; see also Jonathan Schell, *The Fate of the Earth* (New York: Knopf, 1982).

114. See symposium on "Nuclear Winter" in *Issues in Science and Technology,* Winter 1985.

115. Schell, *The Abolition,* p. 22.

116. Parfit argues that a war that killed 100 percent of the world's population would be much worse than one that

killed 99 percent. Derek Parfit, *Reasons and Persons* (New York: Oxford University Press, 1984), p. 453.

117. See the essays on obligations to future generations by Brian Barry, Derek Parfit, and Talbot Page in Douglas MacLean and Peter Brown, eds., *Energy and the Future* (Totowa, N.J.: Rowman & Allanheld, 1983). Also R. I. Sikora and Brian Barry, eds., *Obligations to Future Generations* (Philadelphia: Temple University Press, 1978), and the review by Jefferson McMahon in *Ethics*, Vol. 92, October 1981; Gregory Kavka, "The Paradox of Future Individuals," and Derek Parfit, "Future Generations: Further Problems" *Philosophy and Public Affairs*, Vol. 11, No. 2 (1981); and Bernard Williams, *Ethics and the Limits of Philosophy* (Cambridge: Harvard University Press, 1985), Chapter 9.

118. More formally specified, the formula is: If $P$ $(F)$ is the probability of failure, and $n$ the number of trials, the likelihood of at least one success is $1 - P(F)^n$

119. Bradford Lyttle, *The Flaw in Deterrence* (Chicago: Midwest Pacifist Publishing Center, 1983), pp. 8–11. Note that Lyttle's numbers depend on the number of missiles that could be accidentally launched. Nuclear war need not follow.

120. I am indebted to James Blight for this point and for helpful conversations on this section in general.

121. Note Helen Caldicott's statement that the buildup of weapons "will make nuclear war a mathematical certainty." *The Defense Monitor*, 13, No. 8 (1984): 8.

122. See "Arms Races and the Causes of War, 1850–1945," in Paul Kennedy, *Strategy and Diplomacy* (London: George Allen & Unwin, 1983), and Samuel P. Huntington, "Arms Races: Prerequisites and Results," *Public Policy*, 1958, pp. 41–83.

123. Frederick Mosteller, Robert Rourke, and George Thomas, Jr., *Probability with Statistical Applications* (Reading, Mass.: Addison-Wesley, 1961), p. 17.

124. Schroeder, "Does Murphy's Law Apply?" (note 109), pp.

86, 93. Contrast this with the bleak view of a "senseless future" which Susan Khin Zaw uses to justify an emotional approach to nuclear ethics. Zaw, "Morality and Survival" (note 19).

125. Robert E. Goodin, "Nuclear Disarmament as a Moral Certainty," *Ethics*, April 1985, p. 644.

126. H. W. Lewis, "Technological Risk," unpublished paper.

127. For example, Paul Slovic, Baruch Fischoff, and Sarah Lichtenstein, "Facts and Fears: Understanding Perceived Risk," in Richard Schwing and Walter Albers, eds., *Societal Risk Assessment* (New York: Plenum Press, 1980).

128. Hardin, "Risking Armageddon" (note 106).

129. Lackey, *Moral Principles and Nuclear Weapons* (note 41).

130. Bernard Williams, "Morality, Scepticism and the Nuclear Arms Race," in Blake and Pole, *Objections to Nuclear Defence* (note 16), pp. 108–9.

131. McGeorge Bundy, "To Cap the Volcano," *Foreign Affairs*, Vol 48, October 1969. Henry Kissinger reports of his time in office, "I did not believe at any time that we were close to nuclear war." *Los Angeles Times*, August 11, 1985.

132. Michael Howard, *The Causes of Wars* (London: Unwin Paperbacks, 1983).

133. Paul Bracken, "Accidental Nuclear War," in Allison, Carnesale, and Nye, *Hawks, Doves, and Owls* (note 91).

134. *Ibid.*, Chapter 8, for detailed argument and evidence.

135. See Michael MccGwire, "The Dilemmas and Delusions of Deterrence," in Gwyn Prins, ed., *The Nuclear Crisis Reader* (New York: Vintage, 1984), p. 78. James Blight has also used the concept of reassurance in unpublished papers shared with me.

136. Robert Kennedy, *Thirteen Days* (note 9). I am also indebted to McGeorge Bundy for clarifying this description of President Kennedy's assessment of the situation during the crisis.

137. Lackey, *Moral Principles and Nuclear Weapons*, pp. 26, 154–55.

138. On the problems of how to improve cost-benefit calculations, see Herman Leonard and Richard Zeckhauser, "Cost-Benefit Analysis Applied to Risks," in Douglas MacLean, ed., *Values at Risk* (Totowa, N.J.: Rowman & Allanheld, 1985).

139. Lifton and Falk, *Indefensible Weapons* (note 18).

140. Walzer, *Just and Unjust War* (note 3), Ch. 17.

141. *New York Times*, October 16, 1984, p. 16, and Robert Coles, "The Doomsayers," *Boston Observer*, Vol. 3, November 1984.

142. Kaplan, *Wizards of Armaggedon* (note 7).

143. Lifton and Falk, *Indefensible Weapons*, p. 262.

144. Michael Walzer, "Deterrence and Democracy," *The New Republic*, July 2, 1984, p. 17. See also the discussion in Robert Dahl, *Controlling Nuclear Weapons: Democracy Versus Guardianship* (Syracuse: Syracuse University Press, 1985).

145. Javier Perez de Cuellar, "Secretary General's Statement to General Assembly on Disarmament Issues" (New York: United Nations December SG/SM/3635, 12 December 1984).

146. Quoted in *New York Times*, April 26, 1985, p. 10.

147. Lackey, *Moral Principles and Nuclear Weapons*, p. 187.

148. *Ibid*, p. 157.

149. Rose Mary Volbrecht, "Nuclear Deterrence: Moral Dilemmas and Risks," *Philosophy and Social Criticism*, 10, No. 3/4 (1984): 139–40.

150. I am indebted to David Rosenberg and Ernest May for discussion of this point. See also Richard Ned Lebow, "Windows of Opportunity: Do States Jump Through Them?," *International Security*, 9 (Summer 1984), p. 173.

151. See Harvard Nuclear Study Group, *Living With Nuclear Weapons* (note 79), Chapter 8, for details.

152. Kenneth Waltz, "The Spread of Nuclear Weapons: More May Be Better" (London: International Institute of Strategic Studies Adelphi Paper 171, 1981).

153. See Harry Rowen, "Catalytic Nuclear War," in Allison, Carnesale, and Nye, *Hawks, Doves, and Owls* (note 91).

154. Bernard Williams, in Blake and Pole, *Objections to Nuclear Defence* (note 16), p. 109.

155. See the discussion in Lawrence D. Freedman, "Indignation, Influence and Strategic Studies," Inaugural lecture in the Chair of War Studies, Kings College, London, November 22, 1983.

156. For example, see the critique by James Stegenga, "Nuclear Deterrence: Bankrupt Ideology," *Policy Sciences*, 16 (1983): 127–45.

157. Hollenbach, "Ethics in Distress" (note 37).

158. Schall, *Out of Justice, Peace* (note 75), and U.S. Catholic Conference, "Challenge of Peace" (note 1), p. 18.

159. For details, see Allison, Carnesale, and Nye, *Hawks, Doves, and Owls*, Chapter 9.

160. For an empirical test, see Paul Huth and Bruce Russett, "What Makes Deterrence Work? Cases from 1900 to 1980," *World Politics*, 36 (July 1984): 496–526.

161. For a framing of the debate over "no first use," see McGeorge Bundy *et al.*, "Nuclear Weapons and the Atlantic Alliance," *Foreign Affairs*, Vol. 60, Spring 1982, and the reply by Karl Kaiser *et al.*, "Nuclear Weapons and the Preservation of Peace," *Foreign Affairs*, 60, Summer 1982. Senator Sam Nunn has made the sensible suggestion that any declaration on no first use could be made contingent on Soviet pullbacks of some of the armored divisions that cause NATO to wish to retain the option of nuclear use against Soviet conventional attack.

162. French bishops in Schall, *Out of Justice, Peace*, p. 103.

163. U.S. Catholic Conference, "Challenge of Peace," p. 18.

164. McGeorge Bundy, "Extential Deterrence and Its Consequences," in Douglas Maclean, ed., *The Security Gamble* (Totowa, N.J.: Rowman & Allanheld, 1984).

165. Hardin, "Deterrence and Moral Theory" (note 31).

166. This is pointed out by Charles Glaser in "Why Strategists Disagree About the Requirements of Deterrence," un-

published paper. William Arkin and Richard Fieldhouse claim that under current plans "in each of the 200 largest Soviet cities, an average of 19.1 warheads with 6.33 equivalent megatons would be exploded." *Nuclear Battlefields* (Cambridge: Ballinger, 1985), p. 94.

167. For a recent defense of MAD, see Lee, "Morality of Nuclear Deterrence" (note 40), and Paul Kattenburg, "MAD is the Moral Position," in Charles Kegley and Eugene Wittkopf, eds., *The Nuclear Reader* (New York: St. Martin's Press, 1985).

168. See Kaplan, *Wizards of Armageddon* (note 7); Leon Wieseltier, *Nuclear War, Nuclear Peace* (New York: Holt, Rinehart & Winston, 1983); and Albert Wohlstetter, "Bishops, Statesmen, and Other Strategists on the Bombing of Innocents," *Commentary*, June 1983.

169. The assumption of only two alternatives is a flaw in Susan Okin's otherwise intelligent criticism of the American bishops. "Taking the Bishops Seriously," *World Politics*, Vol. 36, July 1984.

170. Arthur L. Burns, *Ethics and Deterrence: A Nuclear Balance Without Hostage Cities* (London: Institute of Strategic Studies Adelphi Paper No. 69, July 1970), and Bruce Russett, "A Countercombatant Alternative to Nuclear MADness," in Ford and Winters, *Ethics and Nuclear Strategy* (note 87).

171. Arkin and Fieldhouse say there are some 2,000 strategic force targets (silos, sub bases, airfields, weapons storage); 4,500 communications, radar, and defense targets; and less than 1,000 conventional military force targets among the 40,000 in current U.S. plans. Arkin and Fieldhouse, *Nuclear Battlefields*, p. 93.

172. See Scott Sagan, "The Department of Retaliation," *The New Republic*, June 17, 1985, p. 39.

173. I am indebted to Thomas Schelling for help on this section, though we disagree on whether limited counter-city or limited counter-combatant targeting would be more likely to save innocent children. Schelling argues that if tactical nuclear war has broken out, a city targeting capa-

bility is meant to be there to deter further unlimited escalation.

174. Leon Wieseltier argues that a strike against cities "will almost certainly preclude the possibility of any further strikes" and that war termination should be our primary goal if deterrence fails. "When Deterrence Fails," *Foreign Affairs*, 63 (Spring 1985): 841.

175. This is a deficiency in the debate between Albert Wohlstetter and his critics in *Commentary*, cited in note 168. See also Pierre Hassner, "Arms Control and Morality," in James E. Dougherty *et al.*, eds., *Ethics, Deterrence, and National Security* (Washington, D.C.: Pergamon-Brassey's, 1985).

176. See Seweryn Bialer, "The Psychology of U.S.–Soviet Relations," Gabriel Silver Lecture, Columbia University, 1983; and Richard Pipes, *Survival Is Not Enough* (New York: Simon & Schuster, 1984), for contrasting views.

177. See Steven Kull, "Nuclear Nonsense," *Foreign Policy*, Vol. 58, Spring 1985.

178. Compare Robert Jervis, *The Illogic of American Nuclear Strategy* (Ithaca, N.Y.: Cornell University Press, 1984), and Harold Brown, *Thinking About National Security* (Boulder, Colo.: Westview Press, 1983).

179. See discussion in Woolsey, *Nuclear Arms* (note 74).

180. I owe this phrase to a historian, William H. McNeill.

181. Allison, Carnesale, and Nye, *Hawks, Doves and Owls* (note 91), Chapter 9.

182. Garwin, quoted in "Is There a Way Out?" *Harpers*, June 1985, p. 39.

183. The failure of nuclear deterrence is a more complex concept than it might appear at first glance. In a narrow sense war by accident is not a failure of deterrence, but in a broader sense it is a failure of the system of defense based on nuclear deterrence. In addition, a failure of deterrence need not lead to nuclear war. It can also lead to surrender of important values without war or to conventional or chemical and biological war.

184. Richard Ullman, "Denuclearizing International Politics," *Ethics*, Vol. 95, No. 3 (April 1985).
185. Tucker, "Morality and Deterrence" (note 83).
186. See Michael Doyle, "Kant, Liberal Legacies, and Foreign Affairs," *Philosophy and Public Affairs*, Vol. 12, No. 3 (Summer 1983) and No. 4 (Fall 1984).
187. Albert Wohlstetter, "Between an Unfree World and None," *Foreign Affairs*, 63 (Summer 1985): 989–90.
188. For an introduction to the debate, see the articles by James Fletcher, George Keyworth, Sidney Drell, and Wolfgang Panofsky in *Issues in Science and Technology*, Vol. 1, Fall 1984.
189. See Paul Nitze, *On the Road to a More Stable Peace* (Washington, D.C.: State Department, February 20, 1985). It is also worth noting that Freeman Dyson, *Weapons and Hope* (note 15), supports such defense only on the premise that it *follows* offensive reductions.
190. In the words of G. Shakhnazarov, the President of the Soviet Political Science Association, "The Marxist-Leninist conception has never opposed the class origin to the human one. On the historical scale, the all human is eternal while the class is transient. . . . Nobody needs an ideal social system if it has been obtained by such a monstrous price . . . There are no political aims which can justify the use of means which can bring a nuclear war." "Logic of the Nuclear Era," *Social Sciences* (U.S.S.R. Academy of Sciences), 26, No. 2 (1985): 47.
191. Stanley Hoffmann, *Primacy or World Order* (New York: McGraw-Hill, 1978), p. 11.
192. See Robert Keohane, *After Hegemony* (Princeton, N.J.: Princeton University Press, 1984).
193. Robert Axelrod, *The Evolution of Cooperation*, (New York: Basic Books, 1984).
194. See J. S. Nye, ed., *The Making of America's Soviet Policy.* (New Haven: Yale University Press, 1984).
195. See Robert Keohane and Joseph S. Nye, Jr., eds., *Transnational Relations and World Politics* (Cambridge, Mass.: Harvard University Press, 1972).

196. See J. S. Nye, *Peace in Parts* (Boston: Little Brown, 1971), for a discussion of federalist, functionalist and neofunctionalist strategies of building peace. See also Karl W. Deutsch *et al.*, *Political Community in the North Atlantic Area* (Princeton, N.J.: Princeton University Press, 1957), for a discussion of "pluralistic security communities," where states have reasonable expectations of nonviolent behavior.

# For Further Reading

Even the interest of a serious reader can be quickly drowned in the tide of books and articles about nuclear issues. A list for further reading must be brief to be useful, but brevity requires arbitrary decisions that leave out worthy works. Part of that problem can be remedied by reading the footnotes, which provide a trail for deeper intellectual pursuit. Similarly, while this list emphasizes more recent books, they in turn provide a path to the past for those who wish to pursue it.

## Moral Philosophy

Tom L. Beauchamp. *Philosophical Ethics: An Introduction to Moral Philosophy.* New York: McGraw-Hill, 1982. A useful introductory reader.

William K. Frankena. *Ethics.* Englewood Cliffs, N.J.: Prentice-Hall, 1973. A brief survey of the field.

J. J. C. Smart and Bernard Williams. *Utilitarianism: For and Against.* Cambridge: Cambridge University Press, 1973. An illuminating debate between two fine philosophers.

Alasdair MacIntyre. *After Virtue.* Notre Dame, Ind.: University of Notre Dame Press, 1981.

Bernard Williams. *Ethics and the Limits of Philosophy.* Cambridge: Harvard University Press, 1985.

Two difficult but fascinating overviews of the field.

## Ethics and International Politics

Charles Beitz. *Political Theory and International Relations.* Princeton, N.J.: Princeton University Press, 1979.

J. E. Hare and Carey B. Joynt. *Ethics and International Affairs.* New York: St. Martins, 1982.

Stanley Hoffmann. *Duties Beyond Borders.* (Syracuse, N.Y.: Syracuse University Press, 1981).

Three well-written overviews of a difficult relationship.

Kenneth Thompson, ed. *Moral Dimensions of American Foreign Policy.* (New Brunswick, N.J.: Transaction Books, 1984).

Ernest W. Lefever, ed. *Ethics and World Politics.* Baltimore: Johns Hopkins Press, 1972.

Good collections of essays from the past two decades.

*Ethics* and *Philosophy and Public Affairs* are philosophical journals that frequently carry articles on this subject.

## Just War Doctrine and the Use of Force

Robert E. Osgood and Robert W. Tucker. *Force, Order and Justice.* Baltimore: Johns Hopkins Press, 1967. An intelligent, broad discussion that remains relevant.

Ralph B. Potter. *War and Moral Discourse.* (Richmond, Va.: John Knox Press, 1973). A short, clear statement with a useful bibliographical essay.

James T. Johnson. *Can Modern War Be Just?* New Haven: Yale University Press, 1984. An authority on just war doctrine applies it to current issues.

Michael Walzer. *Just and Unjust Wars.* New York: Basic Books, 1977. This important argument is a modern classic.

## The Nuclear Issue: Recent Overviews

Harvard Nuclear Study Group, *Living with Nuclear Weapons.* Cambridge: Harvard University Press, 1983.

Michael Mandelbaum. *The Nuclear Future.* (Ithaca, N.Y.: Cornell University Press, 1983).

Michael Nacht. *The Age of Vulnerability.* Washington, D.C.: Brookings, 1985.

Leon Wieseltier. *Nuclear War, Nuclear Peace.* (New York: Holt, Rinehart & Winston, 1983).

Four efforts at describing the broad dimensions of the nuclear dilemma in a brief and clear form that is readily accessible to the public.

Jonathan Schell. *The Fate of the Earth.* New York: Knopf, 1982.

———. *The Abolition.* New York: Knopf, 1984. A beautifully written presentation of the abolitionist case.

Freeman Dyson. *Weapons and Hope.* New York: Harper & Row, 1984. An idiosyncratic but interesting overview by an eminent physicist.

*Foreign Affairs, Foreign Policy, and International Security* are journals that frequently treat current issues in the relationship between nuclear weapons and foreign policy.

## Nuclear Ethics

Paul Ramsey. *The Just War.* New York: Scribner's, 1968. A powerful classic in the earlier debate by a leading Protestant thinker.

Judith A. Dwyer, ed. *The Catholic Bishops and Nuclear War.* Washington, D.C.: Georgetown University Press, 1984.

# For Further Reading

David Hollenbach, S.J. *Nuclear Ethics: A Christian Moral Argument.* New York: Paulist Press, 1983.

National Conference of Catholic Bishops, "The Challenge of Peace: God's Promise and Our Response", *Origins* (National Catholic Documentary Service), Vol. 13, May 19, 1983.

Michael Novak. *Moral Clarity in the Nuclear Age.* (Nashville: Thomas Nelson, 1983).

James V. Schall, S.J., ed. *Bishops' Pastoral Letters.* San Francisco: Ignatius Press, 1984.

> Six important presentations of different positions within the Catholic Church on nuclear ethics.

Douglas Lackey. *Moral Principles and Nuclear Weapons.* Totowa, N.J.: Rowman & Allanheld, 1984. A utilitarian philosopher's case for disarmament; provocative and interesting even when not convincing.

Nigel Blake and Kay Pole, eds. *Objections to Nuclear Defence.* London: Routledge & Kegan Paul, 1984.

James E. Dougherty *et al. Ethics, Deterrence and National Security.* Washington, D.C.: Pergamon-Brassey's, 1985.

Geoffrey Goodwin, ed. *Ethics and Nuclear Deterrence.* London: Groom, Held, 1982.

*Ethics,* Vol 95 (April 1985). Special issue on ethics and nuclear deterrence edited by Russell Hardin, John J. Mearsheimer, Gerald Dworkin, and Robert E. Goodin.

Ernest W. Lefever and E. Stephen Hunt, eds. *The Apocalyptic Premise.* Washington, D.C.: Ethics and Public Policy Center, 1982.

Douglas MacLean, ed. *The Security Gamble.* Totowa, N.J.: Rowman & Allanheld, 1985.

James P. Sterba, ed. *The Ethics of War and Nuclear Deterrence.* Belmont, Calif.: Wadsworth, 1985.

R. James Woolsey, ed. *Nuclear Arms: Ethics, Strategy, Politics.* San Francisco, ICS Press, 1984.

> Seven among the numerous collections of diverse sets of essays, often ranging from the trivial to the profound within a single set of covers.

# Index

# Index

Benn, S. I., 135n.
Bennett, Jonathan, 143n.
Bentham, Jeremy, 16
Beres, Rene Louis, 134n.
Berlin crisis, 66
Berlin, Germany, 76
Bialer, Seweryn, 149n.
Biological warfare, 63, 149n.
Bismarck, Otto von, 106, 128
Blair, Bruce G., 142n.
Blight, James, 144n., 145n.
Bok, Sissela, 137n.
Bolshevik Revolution, 127
Bracken, Paul, 145n.
Brandt, Willy, 82
Brazil, x, 82
  as nuclear threshold country, 87
Brezhnev, Leonid, 128

Britain
  as nuclear weapons state, 85
  war with Argentina, 86
Brown, Harold, 149n.
Brown, Peter, 144n.
Bundy, McGeorge, 72, 145n., 147n.
Burns, Arthur L., 148n.

Caldicott, Helen, 144n.
Carnesale, Albert, 142n., 145n., 147n., 150n.
Catholic Bishops: see American Catholic Bishops, French Catholic Bishops, German Catholic Bishops
China, as nuclear weapons state, 85
Chinese-American relations, 29, 66, 127
Cohen, Marshall, 134n.
Colby, William, 134n.
Cold War, 129
Coles, Robert, 77, 146n.
Command, control, and communication, 51, 73
Community, 34, 36
  multiple layers of, 38
Consequentialism, 10–20
  antinuclear, 60, 62–70
  broad, 25–26
  criteria for sound reasoning based on, 80
  limits, 19
  pro-deterrence, 60, 62, 70–77

Consequentialism—*Continued*
  shallow, 23
  uncertainties in, 92
Consequentialists
  antinuclear, 79–80
  and nuclear weapons as means, 49
  and strategic interaction, 54
Conventional forces, relation to nuclear forces, 94–95
Conventional weapons, technological improvements, 124
Cooper, Richard, 139n.
Cosmopolitan approach, 32–34, 35t.
Cosmopolitan-realist synthesis, 34–41
Cosmopolitans, sophisticated, 34
Crusades, 46, 47
Crystal ball effect, 61, 87
Cuba, 118
Cuban missile crisis, 3, 5, 66, 75–76
Cynics, 4–5
Czechoslovakia, 21

Dahl, Robert, 146n.
Decapitation, 118
Detente, 127
Deterrence
  balanced, 119
  beyond (political and social paths), 126–31
  beyond (technological paths), 124–26
  bluff, 53
  calculated, 117
  of conventional war, 53, 103
  dimensions of (inherent and calculated), 107
  as an end, 44
  and ethics of virtue, 17
  existential, 106–107, 111, 113–14
  extended, 101
  finite, 109
  moral considerations, xi, 2–3
  as a moral evil, 50
  opposition to, 11–12
  as self-defense, 47
  as a step toward some indefinite long term, 98
  threats to third countries, 55–57
  weaponless, 94, 107
Dien Bien Phu, 5
Discrimination, 57
Dominican Republic, 21

157

# Index

Impartiality in moral reasoning, 23–24, 83
  distinguished from egalitarianism, 37
India, x, 85, 87
Indonesia, 82
International Atomic Energy Agency, 87
International law, 31, 46
International politics
  and community, 34, 36
  cynic's view of, 4
  ethical considerations, 7–10
  order versus justice in, 29–30
  role of nuclear weapons in, 130–31
  skeptic's view of, 6–7
Iran, 39, 118
Iraq, 9
Islamic "jihad," 43
Israel, 85, 87

Japan, x
Jervis, Robert, 149n.
Johnson, James, 44, 140n., 141n.
Just cause doctrine, 98
  as limited to self-defense, 104
Just defense doctrine, xi, 43
Just deterrence, 132
Just war doctrine, xi, 39, 43–44, 46, 50, 53, 57–58, 92, 98, 108
  and nuclear weapons, 2

Kaiser, Karl, 147n.
Kant, Immanuel, 123
Kantians, 49, 54, 136n.
  and means, 57
  rules, 75
  view of future, 131
Kaplan, Fred, 134n., 143n., 146n., 148n.
Kattenburg, Paul, 148n.
Kavka, Gregory, 58, 142n., 143n., 144n.
Kennedy, John F., 5, 75, 76, 87
Kennedy, Paul, 144
Kennedy, Robert, 134n., 145n.
Keohane, Robert O., 139n., 150n.
Kerckhove, Derrick de, 135
Keyworth, George, 150n.
Khrushchev, Nikita, 76
Kissinger, Henry, 29

Krauthammer, Charles, 133n., 140n.
Kull, Steven, 149n.
Kunkel, Joseph C., 141n.
Kyoto, Japan, 5

Lackey, Douglas, 66, 69–70, 76, 83, 138n., 140n., 145n., 146n.
Launch on warning policy, 118
Lebow, Ned, 146n.
Lee, Steven, 137n., 148n.
Leonard, Herman, 145n.
Lewis, H. W., 145n.
Libya, 95
Lichtenstein, Sarah, 145n.
Lifton, Robert J., 134n., 143n.
Limited nuclear war, 51
Luban, David, 32, 138n.
Luttwak, Edward, 142n.
Lyttle, Bradford, 66, 144n.

MacIntyre, Alasdair, 135n., 139n.
MacLean, Douglas, 144n.
Marxism-Leninism, 46, 150n.
Maxims of nuclear ethics
  minimize harm to innocent people, 108–115
  never treat nuclear weapons as normal weapons, 104–108
  reduce reliance on nuclear weapons over time, 120–31
  reduce risks of war in near term, 115–20
  self-defense, 100–104
May, Ernest, 146n.
MccGwire, Michael, 145n.
McCormick, Richard, 143n.
McNamara, Robert, 53, 142n.
McNeill, William, 149n.
Means and nuclear weapons, 49–58
Meyer, Steven, 142n.
Middle East, 3
Middle East War (1973), 128
Mill, John Stuart, 16
Moliere, 14
Moral absolutists
  and nuclear weapons, 10–11
  trap of, 132
Moral considerations, x–xii
  and crusades, xii
  and deterrence, xi, 2–3; see also Deterrence

# Index

Osgood, Robert, 138n.
Owls, 74, 117–20
  potential errors, 119
Oxford debate, Thompson vs. Weinberger, 48

Pacifism, 422–43
  Christian, 43
Page, Talbot, 144n.
Pakistan, x
  as nuclear threshold country, 87
Panofsky, Wolfgang, 150n.
Parfit, Derek, 65, 143n., 144n.
Paskins, Barrie, 140n.
Pauline principle, 84
Peace of Westphalia (1648), 131
Perez de Cuellar, Javier, 81, 146n.
Pipes, Richard, 149n.
Podhoretz, Norman, 21, 137n.
Political effects of nuclear weapons, 78–79
Pope John Paul II, 43
Potter, Ralph, 141n.
Powers, Thomas, 143n.
Probability of nuclear war, 62, 66–68, 70
  trends over time, 72
Process utopianism, 123
Pro-deterrence consequentialists, 60, 62, 70–77
Proliferation, 85–90, 95
  risk of nuclear weapons use by proliferators, 88
Proportionality, 23, 47, 65, 104
  and the legitimacy of nuclear weapons as means, 50
Psychological scarring due to nuclear weapons, 77
Public views
  on nuclear ethics, 2–6
  on nuclear war, effects, 3
  on probability of nuclear war, 71

Ramsey, Paul, 17, 56, 93, 137n., 141n.
Rawls, John, 30, 37, 138n.
  critique of, 139n.
Reagan, Ronald, 48, 78, 125, 141n.
Realists, 28–30, 34, 35t., 40–41
Religious fundamentalists, 4
Roosevelt, Franklin D., 83
Rosenberg, David, 146n.
Rourke, Robert, 144n.

Rowen, Harry, 146n.
Rule-oriented philosophies, 84; see
  also Kantians; Ethics of virtue
Russett, Bruce, 147n., 148n.

Safire, William, 26, 138n.
Sagan, Scott, 148n.
Schall, James V., 140n., 147n.
Schell, Jonathan, 15, 50, 60, 63, 65, 68–69, 93–96, 133n., 141n., 143n.
Schelling, Thomas, 114, 148n.
Schlafly, Phyllis, 15
Schlesinger, James, 9
Schroeder, Paul, 67, 143n., 144n.
Schwing, Richard, 145n.
Self-defense, 43, 55, 83
  as a condition for moral deterrence, 100–104
Self-interest in international politics, 6–7
Shakhnazarov, G., 150n.
Shaw, William H., 137n., 142n.
Shue, Henry, 140n.
Singer, Peter, 139n.
Skeptic, 35t.
  application of moral concepts to individuals, 9–10
  distinguished from realists, 29
  escape, 6–10
  limits of, 10
Slovic, Paul, 145n.
Snow, C. P., 62
South Africa, as nuclear threshold country, 87
Soviet Union
  first strike, 118
  moral views, 9
  as nuclear weapons state, 85
  perceptions, 116–17
  prospects for internal change, 129
  repression in, 46
  Schell's treatment of, 95
  as secretive, 116
  as threat, 44
  see also U.S.-Soviet relations
State Department, x
State moralists, 30–32, 35t.
Stegenga, James, 147n.
Steinbruner, John, 142n.
Stimson, Henry, 5
Strategic Air Command, 56

161